向美國百大企業學
商用英語

HOW TO WRITE
ENGLISH BUSINESS
DOCUMENTS

1. 美國百大企業是什麼？

美國百大企業（企業名單詳見次頁）是指在各種產業及領域組成大型商業集團，並在美國市場創造最高營收的超大型企業。觀察美國百大企業的營收，與其佔美國國內生產毛額（GDP）的比率，以及企業所提供的工作機會，不難看出美國百大企業不僅在美國有舉足輕重的地位，也對全世界有重大影響。

2. 美國百大企業使用的商業英語為何？

百大企業的商務人士主要利用電子郵件、合約、報告書、報價單等英文文書溝通。這類英文文書的重要性在國際社會上越來越顯著，商務人士也必須透過各種英文文書來與外國企業溝通工作事宜。商用英文文書有多種格式及特定詞彙，須清楚知道其書寫次序及理解字詞意義。

3. 為什麼我們必須學商業英文文書的寫作法？

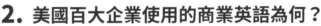

美國百大企業的商務人士撰寫英文文書時會使用特定的格式，本書將說明商務人士使用的各種文書範本，幫助讀者有效率地學習文書撰寫方法、架構及主要表達方式。書中包含 47 種商用英文文書以及 83 個文書範本，只要一書在手，就能乾淨俐落地解決商用英文文書寫作上的疑難雜症。本書收錄了商務上實際使用的文書格式，除了可以直接活用於職場上，對於正在準備就業的讀者也很有幫助。

Top 100 Companies

 2021《財星》雜誌的「財星全球 500 強」選出的美國百大企業

001	Wal-Mart Stores	036	Microsoft	071	Liberty Mutual Insurance Group
002	Exxon Mobil	037	United Technologies	072	Coca-Cola
003	Chevron	038	Dell	073	Humana
004	General Electric	039	Goldman Sachs Group	074	Honeywell International
005	Bank of America Corp	040	Pfizer	075	Abbott Laboratories
006	ConocoPhillips	041	Marathon Oil	076	News Corp
007	AT&T	042	Lowe's	077	HCA
008	Ford Motor	043	United Parcel Service	078	Sunoco
009	J.P. Morgan Chase & Co	044	Lockheed Martin	079	Hess
010	Hewlett-Packard	045	Best Buy	080	Ingram Micro
011	Berkshire Hathaway	046	Dow Chemical	081	Fannie Mae
012	Citigroup	047	Supervalu	082	Time Warner
013	Verizon Communications	048	Sears Holdings	083	Johnson Controls
014	McKesson	049	International Assets Holding	084	Delta Air Lines
015	General Motors	050	PepsiCo	085	Merck
016	American International Group	051	MetLife	086	DuPont
017	Cardinal Health	052	Safeway	087	Tyson Foods
018	CVS Caremark	053	Kraft Foods	088	American Express
019	Wells Fargo	054	Freddie Mac	089	Rite Aid
020	International Business Machines	055	Sysco	090	TIAA-CRER
021	UnitedHealth Group	056	Apple	091	CHS
022	Procter & Gamble	057	Walt Disney	092	Enterprise GP Holdings
023	Kroger	058	Cisco Systems	093	MassInsuranceachusetts Mutual Life
024	AmerisourceBergen	059	Comcast	094	Philip Morris International
025	Costco Wholesale	060	FedEx	095	Raytheon
026	Valero Energy	061	Northrop Grumman	096	Express Scripts
027	Archer Daniels Midland	062	Intel	097	Hartford Financial Services
028	Boeing	063	Aetna	098	Travelers Cos
029	Home Depot	064	New York Life Insurance	099	Publix Super Markets
030	Target	065	Prudential Financial	100	Amazon.com
031	WellPoint	066	Caterpillar		
032	Walgreen	067	Sprint Nextel		
033	Johnson & Johnson	068	Allstate		
034	State Farm Insurance Cos	069	General Dynamics		
035	Medco Health Solutions	070	Morgan Stanley		

百大企業常用的英文文書範例

履歷

DANIEL ALMEIDA

markclark@msn.com | C: 756-7834-5684 | LINKEDIN.com/in/daniel-almeida

Summary

Ambitious and results-driven individual with strong academic credentials and substantial intern experience. Independent professional with high energy and superb communication skills.

Highlights

- Attention to detail
- People oriented
- Multitasker
- Good communicator

Experience

Rogu Ad — Chicago, IL
Ad Sales Intern — Summer 2016
- Worked closely with sales team on organizing strategic plans.
- Coordinated ad campaigns for both radio and print.

MGT Center — Springfield, IL
Sales Intern — Summer 2015
- Interacted with clients in all sales stages.
- Made outbound telephone calls to prospective clients (50 calls a day)
- Arranged orders and delivers for more than 50 clients.

Education

Maruchs College — Chicago, Illinois — 2015-2016
Marketing, — Magna Cum Laude

Activities

- Vice President, Student Bureaucracy Club, 2016
- Treasurer, Beta Gamma Honors Society, 2015-2016

勞動契約

EMPLOYMENT CONTRACT

This Employment Contract (herein "Contract") is made effective as of February 14, 2017, by and between Rocklyn Corporation (herein "Employer") of and Rudy L. Adams (herein "Employee").

1. EMPLOYMENT. Employer hereby employs the Employee as a sales associate for the period beginning March 1, 2017 and ending on the date on which the employment is terminated. Employee agrees to devote fully to the sales affairs of the Employer's products and goods and perform his duties faithfully, industriously and to the best of Employee's ability and experience. Work hours are 40 hours a week.

2. COMPENSATION. As compensation for the services provided by Employee, Employer will pay an annual salary of $30,000 in accordance with payroll procedures.

3. CONFIDENTIALITY. Employee agrees that Employee will not at any time divulge, disclose or communicate any company confidential information to any third party without the prior written consent of Employer. A violation of this will justify legal and/or equitable action by Employer which may include a claim for losses and damages.

4. BENEFITS. Employee shall be entitled to 21 days of paid vacation and 5 days of sick leave per year.

5. TERM/TERMINATION. This Contract may be terminated by Employer upon 1 month written notice, and by Employee upon 1 month written notice.

Laura Swann
Laura Swann, HR Manager
Rocklyn Corporation (EMPLOYER)

Rudy L. Adams
Rudy L. Adams
(EMPLOYEE)

宣傳廣告信

Subject 廣告 James Sutton: You're Invited to Join EZ's Site, WearPlus.com

Dear James Sutton,

We'd like to recommend EZ's private shopping site,

WearPlus.com.

We offer great benefits for your shopping pleasure.

On Sale Every Day: Up to 40% off designer brands, plus great finds for home.

Free Delivery: Free shipping on orders over $69

Powered by EZ: Sign in and make purchases with your existing EZ account.

New event launch daily. Join today!

Unsubscribe | Help

Customer Assistance 1-866-235-5443

©2017 EZ Inc., 1555 Clark Street
Westbury, NY 11590. All rights reserved.

報價單

ESTIMATE

TIEMA

Date: September 24, 2013
Estimate #: 48131311
Valid Until: September 30, 2013
Customer ID: PI719

To   Pineup Corporation
     2303 Liberty Avenue
     Irvine, CA 92618
     714-422-0076

Quantity	Description	Unit Price	Line Total
20	Tiema A4 80 gsm Office Paper (500 sheets)	$9.50	$190.00
2	Basics Stapler (with 1,000 staples)	$4.50	$9.00
		Subtotal	$ 199.00
		VAT Rate %	7.50
		VAT	$ 14.90
		Total	$ 213.90

The above information is not an invoice and only an estimate of services/goods described above. Payment will be collected in prior to provision of services/goods described in this quote.

To accept this quotation, sign here and return : _____

Thank you for your business!
Should you have any inquiries concerning this quote, please contact:
Ronald A. Hensley (530-773-9081)
[Tiema Co., Ltd.] [2484 Byers Lane Redding, CA 96001]
Phone [530-773-9069] Fax [071-114-4932] [info@tiemapaper.com] [www.tiema.com]

內容特色

1. 收錄 47 種實用的商用英文文書

包括美國百大企業間最常見的通用招聘公告、履歷、推薦信等求職相關文書,除此之外還有報告書、會議記錄、宣傳單等基本工作相關文書,再加上各種電子郵件和合約,彙整出共 47 種各式商務英文文書,如實呈現實際的寫作格式。

2. 收錄 83 個商務英文文書範本

就像招聘公告可以分為有經驗者招聘公告和新進員工招聘公告,電子郵件中也有宣傳廣告信和抗議信,我們會遇到各種傳達不同主題和內容的文書。因此書中收錄了多達 83 個商務英文文書範本,讓讀者透過內容各異的文書範本廣泛地應用學習。

3. 百大企業商用英文文書寫作手法

47 種商用英文文書的內容或構成上各有不同的特徵,所以在寫作時必須選對相符的撰寫方式。本書摘錄美國企業間實際使用的文書,並以此為基礎搭配活躍於商場之實務負責人的建議,使讀者能夠更精準地熟練文書寫作方法。

4. 商用英文文書寫作架構

學會寫作方法之後將進入「文書架構詳細解析」單元,幫助讀者看到文書時能夠具體地馬上掌握文章採取何種架構和敘事技巧。文中將收錄的文書分為「標題—引言—本文—結論」等部分,讀者能夠仔細觀察各部分應以什麼樣的內容為重心撰寫。

5. 統整主要用法及詞彙

統整應用於範例中的各種商用文書主要用法及詞彙,可以隨時隨地輕鬆複習。透過搭配的簡單的小測驗,確認自己是否掌握重點內容。

目次 Contents

PART 1
Basic Foundation

奠定基礎

Unit 01 英文寫作基本原則

1. 商用英文文書的 5 項基本原則

❶ 明確的目的

所有的商用文書都要有明確的目的，因商用文書是達成協議、交易、簽約、會議邀請等重要工作的必要手段。因此為了使收件人能夠正確掌握撰寫者的意圖，並因應其意圖做出適當的回應，寫作商用文書時一定要明確表示目的。

❷ 簡單明瞭

撰寫者都希望能透過商用文書傳達特定資訊，為了達到這個目標，文書的撰寫必須以簡單明瞭為原則。一旦文書內容過於冗長，將會降低忙碌上班族的閱讀效率，所以撰寫商用文書時要簡單明瞭地點出核心資訊，使收件人能夠確實了解到重點。

❸ 對收件人的認識

撰寫文書時必須對收件人有明確的認知。正確掌握收件人的身分，才能避免寫出造成誤解或讓人不悅的表達方式或用語，並確實區分出對方須要知道及不必知道的訊息。此外，也要根據對象區分非正式及正式語氣（tone）。

❹ 正確性

撰寫文書時，應該要使用正確的文法、拼寫、結構完整的句子。儘管一兩個錯誤是可以容忍的，但重複出錯也可能會讓收件人對撰寫者失去信任。

❺ 注意格式

商用文書分為提案計畫書、合約、電子郵件等種類，每種文書使用的標準格式各有不同。每間企業或各國的格式可能存在些許差異。不過依照一般標準格式撰寫，將使收件人更容易理解並掌握信件內容，把產生混淆的可能性降到最低。

2. 英文寫作的 16 項原則

❶ 使用簡單的詞彙

有不少人認為使用困難的詞彙寫作能展現較高的英文寫作水準。但有簡單的詞彙可用的話，也不必刻意使用較難的說法。

utilize (X) → use (O)

❷ 避免不必要的詞彙

因為是商用文書，不應加入不必要的用語。

I **have met** you yesterday. (X)

→ I **met** you yesterday. (O)

❸ 避免使用譬喻

盡量避免使用非英文的外文語句、科學性的描述、譬喻式的表達方式如行話（jargon）。如果在商用文書中用到 touch base、think outside the box 等慣用語的話，有可能會造成對方誤解，所以絕對要避免使用。

see eye to eye (X) → agree (O)

❹ 描寫要具體

所謂的具體不是指追求正確性，而是指寫出詳細情形。舉例來說，比起說「目前狀況」，用「公司的財政困難狀態」等的具體表達語句會更好。

❺ 標點符號及大寫的使用

應避免連續使用兩個以上的句點、驚嘆號、問號等，且要避免用大寫寫出整個單字或整句話。因為全部都用大寫的話，對母語人士來說有一種大聲喊叫的感覺，會令人感到不悅。如果有想要強調的內容，建議使用粗體或斜體來表示。

Would you send me the **REPORT**??? (X)

→ Would you send me the **report**? (O)

❻ 被動和主動

比起被動語態，最好使用主動語態。一般來說，使用「主詞—動詞—受詞」的語法結構對讀者而言更易讀。

The meeting **was led by** Tony. (X)

→ Tony **led** the meeting. (O)

❼ 妥善選擇代名詞

選擇代名詞（you、they、me 等）時必須多加注意。以 myself 代替 me 是連英文母語者都常犯的錯誤，句子裡的主詞和受詞相同的話用 myself，主詞和受詞不同則用 me。此外，me 和 I 也經常被混淆使用，要記得當主詞時為 I，當受詞時為 me。

Send the memo to Jane and **myself**. (X)

→ Send the memo to Jane and **me**. (O)

Billy and **me** give you greetings. (X)

→ Billy and **I** give you greetings. (O)

❽ 主詞和動詞單複數相符

要注意句子中的主詞和動詞單複數是否相符。主詞是單數名詞的話動詞要加 s，主詞是複數名詞的話不加 s。順帶一提，當句子中的主詞為 everybody、everyone、nobody、someone、each、either、neither 等時，都要視為單數名詞處理，須在動詞後加上 s。

Everyone **know** where it is. (X)

→ Everyone **knows** where it is. (O)

❾ 冠詞的使用

冠詞是最令人感到棘手的部分之一。冠詞的用法分為置於名詞前的不定冠詞 a(n)、定冠詞 the 以及省略冠詞等三種。只要沒有所有格代名詞（my、you 等）、指示形容詞（this、that 等）或數字（two、three 等）出現，就必須從三種用法中擇一。

冠詞	定義	例句
a(n)	表示「一個」，用來指稱非特定對象的單件事物。僅限用於單數且可數的名詞。	I ate **a bananas**. (X) I drank **a water**. (X) I ate **a banana**. (O)
the	表示「某個」，用來指稱特定對象。可放在單數、複數、可數、不可數名詞之前。	I ate **the banana**. (O) I ate **the bananas**. (O) I drank **the water**. (O)
「省略」	非指特定對象的一般複數名詞可省略冠詞。 非指特定對象的不可數名詞也可省略冠詞。	I ate **banana**. (X) I ate **bananas**. (O) I drank **water**. (O)

❿ **That vs Which**

連接句子和句子的時候，必須知道如何正確使用 that 和 which。That 後面接用來介紹先行詞的「必要資訊」，which 後面則是接用來介紹先行詞的「附加資訊」。

[that + 先行詞的必要資訊]

This is our only product **that** sells well right now.

[which + 先行詞的附加資訊]

I'd like to introduce a new product, **which** has the potential to revolutionize the market.

⓫ Affect vs Effect

Affect 和 effect 是商用文書中的常見單字。Affect 表示「造成影響」，多作動詞使用。而 effect 主要當名詞使用，表示「結果」，要注意它們之間的差別。

[主要當動詞使用的 affect] – The weather **affected** the day's activities.

[主要當名詞使用的 effect] – The new release had a great **effect** on the company's sales.

⓬ He vs She vs They

第三人稱代名詞的性別問題是連英文母語者都感到混亂的部分。近年來由於多元性別認同的議題受到重視，有人開始使用複數形的 they 來代替單數代名詞 he。不過依然有不少人不認同或不適應這種用法，建議乾脆使用 user、person、you 等單字。最好也避免使用「he/she」或「s/he」等字詞。

If a member has enough points, **he** may receive a free gift. (X)

→ **A member** with enough points may receive a free gift. (O)

Right click **her** name and click Select. (X)

→ Right click the **user**'s name and click Select. (O)

⓭引號的位置

英文句子中存在 " " 時，句尾的句號、逗號要放在引號裡面。這個錯誤出現的機率極高，須多加留意。

Many critics call it "business jargon". (X)

→ Many critics call it "business jargon." (O)

⓮數字寫法

0 至 10 以文字表示，11 以上的數字則用阿拉伯數字。不過用於度量單位時，10 以下的數字也一樣用阿拉伯數字表示。當度量單位前面的數字為百萬（million）等概數時，用阿拉伯數字加上 million 表示。此類度量單位包括距離、溫度、體積、尺寸、重量等單位，kb（kilobyte）、mb（megabyte）也包含在內。

3 computers (X) → **three** computers (O)

eight km (X) → **8** km (O)

zero megabytes (X) → **0** megabytes (O)

five million (X) → **5** million (O)

6 **million** 980 **thousand** (X) → **6,980,000** (O)

⓯日期寫法

在美國日期格式順序為月 - 日 - 年，例：January 1, 2017。讀日期的時候，「1（1 日）→ 讀作序數（ordinal number）first」，但在撰寫文書時必須以基數（cardinal number）表示，寫為「January 1」。而且月份不能寫為阿拉伯數字或縮寫。舉例來說，原則上不會將 3 月、4 月寫為「3、4、Mar.、Apr.」，不過有時還是會因為空間不足等原因採取縮寫。此外，歐洲、中南美洲、澳洲等地的日期格式為日 - 月 - 年，這一點要多加注意。

January 1st, 2017 (X) 1/1/2017 (X) Jan. 1, 2017 (X)

January 1, 2017 (O) 12 November, 2013 (O)

⓰時間寫法

在國際商業領域上，最好使用 24 小時制表示時間。例如「下午 3 點→ 15 點」。不過子夜不是寫為 24:00，而是 0:00。美國人常用 A.M.（上午）和 P.M.（下午），原則上是大寫（例：寫為 P.M. 而非 pm），但若想表示上午 12 點或下午 12 點的話，會寫成 12:00 A.M. or 12:00 P.M.。但為避免產生混淆的狀況，建議寫為 noon（正午）或 midnight（子夜）。另外，最好不要在正式商用文書中使用 o'clock（如 1 o'clock），且與他國人士交流時要連同時區標示清楚。

The meeting is at **12:00 P.M.** (X)

→ The meeting is at **noon**. (O)

We will arrive at **1 o'clock in the afternoon**. (X)

→ We will arrive at **13:00 Pacific Time**. (O)

MEMO

Unit 02

常犯的英文寫作錯誤

01. Mr. John

很多人寫電子郵件時，常常在 Mr./Ms. 後面加上名字，而不是姓。一定要加姓才可以。Mr./Ms. 後接 full name（名加姓）雖不常見，但可以使用。

Dear **Mr. John**, (X)

→ Dear **Mr. Reynolds**, (O)

　　Dear **Mr. John Reynolds**, (O)

02. I have a promise/schedule.

許多人用英文提到約會或行程時，會寫成表示「約定」的 promise 或是表示「行程」的 schedule。但要表示「有事、有行程了」時，要用 plans 或 appointment 才對。

I have **a promise** with my boss. (X)

I have **a schedule** on Tuesday. (X)

→ I have **plans** tonight. (O)

　　I have **an appointment** at 8 P.M. (O)

03. Hong Gil-dong

西方人名是先名後姓，所以和外國人溝通時，最好以「名—姓」的格式寫名字。此外，名中間通常會加入連字號（-），不過也可不加，不加時名字的兩個字中間不加空格，否則外國人會誤以為是中間名（middle name）。

Hong Gil-dong (X) Gil Dong Hong (X)

→ Gil-dong Hong (O) Gildong Hong (O)

04. Kim President

以職稱稱呼人的時候會說「金部長」，把職稱放在姓後面。美國也會用職稱稱呼，不過要注意的是美國是先職稱後姓。

Kim President (X) → President Kim (O)

05. I'm David. vs This is David.

在電子郵件中自我介紹時說 I'm David. 或 I am Sumin Kim. 不能說是錯誤的表達方式，但用於寫作上不太自然。寫作時最好用 This is ... 或 My name is ... 表示。

I'm David Choi. (X) → **This is** David Choi. (O)

My name is David Choi. (O)

06. Its vs It's

這是連美國人都常犯的錯誤，相當多人會在應該用 it's 的時候寫成 its。it's 是 it is 的縮寫，its 則是表示「它的」，為所有格代名詞。兩者意思完全不同，因此必須正確區分該不該加省略號。

Its the CEO's private folder. (X)

→ **It's** the CEO's private folder. (O)

07. Whose vs Who's

跟 It's、its 混淆使用的情況類似，whose 和 who's 也經常因為看起來很像，而發生誤用。Whose 是 who 的所有格，who's 則是 who is 的縮寫。兩者意思相差甚遠，一定要多加注意。

Who's contract is it? (X)

→ **Whose** contract is it? (O)

08. Lose vs Loose

Lose 和 loose 看起來很像，所以寫作時經常會誤用。Lose 有「失去」、「輸」等意思，loose 則是表示「鬆開的」為形容詞。 差「o」一個字母，意義就相差甚遠，因此在寫作時一定要小心。

Do not **loose** the keys. (X)

→ Do not **lose** the keys. (O)

09. It was nice meeting you.

很多人在表達見到某人很開心時，會有混淆下述兩種說法的傾向。初次見到某人的時候要用 meeting，見到認識的人時則要用 seeing。

[初次見面時] – It was nice **meeting** you.

[見到認識的人時] – It was nice **seeing** you.

10. Conference vs Meeting

許多人把 conference 當成「會議」，因而用 conference 來代替 meeting。Meeting 主要指在公司內部會議室，和其他職員或客戶開的小型會議，而 conference 則是指規模較大、較正式的活動。

Why don't we have **a conference** at my office tomorrow? (X)

→ Why don't we have **a meeting** at my office tomorrow? (O)

11. Juniors、Seniors

許多人在職場上會用 junior 和 senior 來稱呼前後輩，不過在美國 junior/senior 不是用來稱呼前輩/後輩，而是用來表示職責/職級，例如 junior engineer、senior analyst 等。如果想要稱呼後輩、前輩的話，可以說 junior/senior colleague。

I had a drink with my **juniors**. (X)

→ I had a drink with my **junior colleagues**. (O)

12. Expect

常有人誤以為 look forward to 和 expect 的意思相同，但 look forward to 是表示「期待」某件事，而 expect 是表示「預期」某件事將會發生，使用時必須留意兩者之間的區別。

I'm **expecting** to meet you. (X)

→ I'm **looking forward to** meeting you. (O)

13. Eat dinner

用「eat dinner」表示「吃晚餐」的話，在語意上並沒有問題，不過在商用文書中使用 eat 會顯得過於白話。用 have dinner、have breakfast 等說法來表達較為正式有禮。

We are scheduled to **eat dinner** at 6 P.M. (X) → We are scheduled to **have dinner** at 6 P.M. (O)

14. I have many works to do.

英文須區分可數和不可數名詞，使用時一定要留意這一點。尤其是容易搞混的名詞，例如 work、research、evidence、equipment、truth 等。描述不可數名詞的數量時要用 much、little，不可用 many、few。

I have **many works** to do. (X)

→ I have **much work** to do. (O)

15. Can I...?

can 或 may 可以用來表示請求允許，不過 can 適用於詢問身體上的能力，而電子郵件等商用文書中則應使用 may。

Can I visit your office next week? (X)

→ **May** I visit your office next week? (O)

16. ASAP

熟識的朋友之間使用縮寫沒問題，然而正式的商用文書中必須避免使用 ASAP（as soon as possible）、FYI（for your information）、BTW（by the way）、ATTN（attention）、N/A（not applicable）、e.g.（例如）等縮寫。縮寫不但看起來沒誠意，還會使對方搞混實際用意。

We need the press release **ASAP**. (X)

→ We need the press release **as soon as possible**. (O)

17. I received it well.

很多人會把「我順利收到信了」直接翻成「I received it well.」，此時須省略 well，直接寫 I received it. 就好。

I **received** the package **well** today. (X)

→ I **received** the package today. (O)

18. a(n) vs the

Unit 1 有提到冠詞的使用規則，但因漏掉冠詞的狀況經常發生，所以每當遇到名詞，須確認是否有加 a(n) 或 the。

I used **computer**. (X)

[使用任一台電腦時] – I used **a computer**.

[使用特定的某一台電腦時] – I used **the computer**.

19. I'm doing good.

當有人問 How are you? 的時候，我們常會回應 I'm doing good.。不過這句話為口語表達用法，撰寫商用文書時應避免使用。

The company is **doing good**. (X)

→ The company is **doing well**. (O)

20. Notebook? Laptop?

在美國只會用 laptop 稱呼筆記型電腦，不過幾年前起也有人使用 notebook 來表示筆電。不同製造商間使用的用語各有不同，舉例來說，蘋果公司的筆電是 Macbook，HP 的筆電是 laptop，一般消費者不會去區分兩種用語間的差異。不過如果想避免和紙本筆記本產生混淆，可以用 notebook computer 更精準地表達。

We ordered 23 **notebooks**. (X)

→ We ordered 23 **notebook computers**. (O)

We ordered 23 **laptops**. (O)

We ordered 11 desktops and 23 **notebooks**. (O)

21. A.M. 7:30

寫出時間的時候，要避免用「上午 7 點 30 分」的語序思考，把英文直譯成「A.M. 7:30」。英文中的 A.M.（上午）P.M.（下午）要寫在時間之後。

The seminar will be held at **P.M. 3:30.** (X)

→ The seminar will be held at **3:30 P.M.** (O)

22. Everyday vs Every day

Everyday 是用來修飾每天都做的事（名詞）的「形容詞」，every day 則是由形容詞 every 加上名詞 day 組成的副詞，用來修飾每天都做的活動（動詞）。

I go to work **everyday**. (X)

→ I go to work **every day**. (O)

These are my **everyday** activities. (O)

23. 括號前要加空格嗎？

英文中括號前必須加空格。

It must be generated from Team Foundation **Server(TFS)**. (X)

→ It must be generated from Team Foundation **Server (TFS)**. (O)

24. My vs Our

美國人在所有格的使用上比起 our 更常用 my。因此 my boss、my company 會比 our boss、our company 更恰當。

Our boss will also be there. (X)

→ **My boss** will also be there. (O)

25. 時間要用 on 還是 in ？

原則上日期、星期幾或特定的日子前加 on，月份、年份前加 in。

[用 on 時] – Mr. Jenkins will arrive **on** August 31.

[用 in 時] – They will inspect the factory **in** December.

26. Especially

副詞 especially 不可放在句首，especially 要放在動詞前、形容詞前或是句尾。

Especially, young drivers like fast cars. (X)

→ Young drivers **especially** like fast cars. (O)

Young drivers like fast cars **especially**. (O)

27. Spelling

有些困難的詞語連英文母語者都不太熟悉，一不小心就會拼錯。下表整理出商用文書的常用詞彙，寫作時必須小心，避免打錯。

正確拼法	常見錯字
accommodate	accomodate, acommodate
column	colum
committed	commited, comitted
committee	commitee
consensus	concensus
disappoint	disapoint
embarrass	embarass
immediately	imediately
independent	independant
interrupt	interupt
foreign	foriegn
forty	fourty
grateful	gratefull
guarantee	garantee
leisure	liesure
license	lisence
misspell	mispell, misspel
necessary	neccessary
occasionally	occasionaly, occassionally
occurred	occured
omission	ommision, omision personell,
personnel	personel
plagiarize	plagerize
pronunciation	pronounciation
receipt	reciept
separate	seperate
successful	succesful
questionnaire	questionaire, questionnair

PART 2
Employment Documents

求職文書

招聘公告

 文書範本 1 有經驗者招聘公告

▶翻譯請參閱 p.198

Web Services Developer
NumberOne Soft
Austin, Texas

NumberOne Soft is seeking a senior-level web developer with a focus on front-end and user experience development.

Responsibilities

- Work closely with other developers in UX & UI design, product management and engineering to develop complicated user applications using core web technologies such as HTML, CSS, and JavaScript.
- Constantly upgrade front-end systems to enable rapid UI development and iteration.
- Document processes and workflow in sharable documents for future adoption and reuse.

Basic Qualifications

- Minimum of 2 years' experience reviewing and writing front-end and back-end code.
- An expert-level grasp of HTML, CSS, and JavaScript.
- Expertise in at least one back-end language such as Node.js, Ruby on Rails or PHP.
- Strong comfort working with database programs like MongoDB or MySQL.
- Experience in building, maintaining, and supporting server environments with dedicated hosting providers.

Preferred Qualifications

- A mobile-first approach to web design.
- A habit of browser testing all of your output.
- Empathy for the users of your products and anticipation of their needs.
- Familiarity with design tools like Photoshop, Illustrator, and Sketch.

單字

- seek 尋求、尋找
- web developer 網頁開發工程師
- responsibility 責任、工作
- rapid 迅速的、快速的
- iteration 疊代、重複
- document 記錄 (動詞)
- dedicated 專用的
- preferred 偏好的
- habit 習慣
- empathy 同理心
- anticipation 預想、預測
- familiarity 熟悉、熟練

美國百大企業文書撰寫祕訣！

Job Posting 是為了招聘職員而發布的招聘公告。觀察美國大企業的招聘公告文件，會發現內容都以「應徵條件」為中心，較少提到工作環境及福利。外商特別看重「合適的經歷」，不太要求特別的出身背景、工作態度或學歷，而對於初階職員或實習生的招聘公告會更講求「技術或人格特質」。代表性的國際徵才網站有 Indeed.com、LinkedIn.com 等。

實務負責人的一句話！

要以負責的業務和求職條件為重點仔細撰寫。避免要求太多與經歷無關的資訊，例如年紀和性別等。運用項目符號標示，使文書簡明易讀。

文書架構詳細解析

① Web Services Developer
NumberOne Soft
Austin, Texas

② NumberOne Soft is seeking a senior-level web developer with a focus on front-end and user experience development.

Responsibilities
③
- Work closely with other developers in UX & UI design, product management and engineering to develop complicated user applications using core web technologies such as HTML, CSS, and JavaScript.
- Constantly upgrade front-end systems to enable rapid UI development and iteration.
- Document processes and workflow in sharable documents for future adoption and reuse.

Basic Qualifications
④
- Minimum of 2 years' experience reviewing and writing front-end and back-end code.
- An expert-level grasp of HTML, CSS, and JavaScript.
- Expertise in at least one back-end language such as Node.js, Ruby on Rails or PHP.
- Strong comfort working with database programs like MongoDB or MySQL.
- Experience in building, maintaining, and supporting server environments with dedicated hosting providers.

Preferred Qualifications
⑤
- A mobile-first approach to web design.
- A habit of browser testing all of your output.
- Empathy for the users of your products and anticipation of their needs.
- Familiarity with design tools like Photoshop, Illustrator, and Sketch.

❶ 標題（Heading）
最上方標明職務、公司名稱、工作條件（地區、招聘型態等）。

❷ 職務說明（Job Description）
簡略說明該職務並記載主要事項。

❸ 負責業務（Responsibilities）
具體列出業務內容。通常針對有經驗者的招聘公告，此部分會寫得較為詳細。

❹ 基本條件（Basic Qualifications）
列出應徵資格的重要項目，和業務內容一樣須詳細撰寫。

❺ 加分條件（Preferred Qualifications）
列出公司在招聘人才時格外偏好的非必要條件。可於加分條件下方的「Education」項目進一步記載及要求提供的學歷事項。

撰寫文書所需的主要用語

01 be seeking... 在尋找…
Startup Enterprise **is seeking** an entry level accountant.
新創公司正在尋找初階會計師。

02 (have) expert-level grasp of... 對於…的專業知識
Expert-level grasp of PHP and Perl.
對於 PHP 及 Perl 的專業知識。

03 (have) experience in/with... 對於…的經驗
Experience in working with business associates and consultants.
與商業夥伴及顧問合作的經驗。

04 this is a unique opportunity to... 這是能夠…的特別機會。
This is a unique opportunity to develop leadership skills.
這是能夠開發領導能力的特別機會。

05 be exposed to... 暴露於…、接觸到…
You'll **be exposed to** a broad range of actuarial work.
你將接觸到廣範圍的精算工作。

06 assist with/in... 協助…
Assist with organizing analytical data.
協助整理分析資料。

07 collaborate with/in/on... 與…合作（with 用於人 / 公司，in/on 用於工作）
Collaborate with sales strategists on marketing campaigns.
和銷售策略顧問合作行銷活動。

08 (have) familiarity with... 對…熟悉
Familiarity with CAD tools such as AutoCAD 360.
熟悉 AutoCAD 360 等 CAD 工具。

小測驗

01. KP Group (   ) an experienced radio/TV technician.
    KP 集團正在尋找有經驗的廣播 / 電視技術人員。

02. (   ) with InDesign and Photoshop.
    熟悉 InDesign 和 Photoshop。

03. You'll (   ) to a broad range of marketing opportunities.
    你將接觸到廣泛的行銷機會。

04. This is a (   ) to work with top medical personnel.
    這是能夠和頂尖的醫療人員工作的特別機會。

Actuarial Analyst

Calloway Insurance
New York, New York

Employment type: Full-time
Experience: Entry level

Position Overview

Calloway Insurance is seeking an entry level actuarial analyst for our New York City location. This is a unique opportunity to be exposed to a broad range of actuarial work in one of the leading companies in the field.

Job Functions

- Conduct pricing analysis of individual accounts.
- Analyze data and calculate the probability of costs associated with events such as accidents, property damage, injury and death.
- Produce reports covering underlying data.
- Assist with organizing actuarial data and analytics to enhance processes and maintain a high standard of data integrity.

Requirements

- Bachelor's degree or equivalent training in actuarial science, mathematics, or statistics
- 0 to 2 years of related experience
- One or more actuarial exams passed
- Intermediate to advanced Excel skills
- SQL proficiency preferred
- Excellent problem solving skills
- Solid verbal and written communication skills

Company Description

Calloway started operations in 1921 in the field of insurance investment and now employs more than 11,000 associates worldwide.

單字

- actuarial 保險精算的
- leading 一流、領先的
- pricing analysis 定價分析
- underlying data 基層數據
- enhance 加強
- data integrity 數據完整性
- equivalent training 同等訓練
- proficiency 熟練
- problem solving skills 問題解決能力
- investment 投資
- employ 雇用
- worldwide 在全世界

履歷

 文書範本 1 有經驗者的履歷

▶翻譯請參閱 p.200

Mark C. Clark

2220 Conner St. Gulfport, FL 39501 | Cell: (494) 7834-5684 | markclark@msn.com

WEB CONTENT MANAGER

Senior content manager with 15 years of online marketing experience. Managed more than 150 websites for various companies.

- Growth strategies
- Multimedia experience
- Proactive e-commerce
- Superior collaboration
- Cross-media expertise
- Customer service oriented

PROFESSIONAL EXPERIENCE

Webgem Corporation   **Jacksonville, FL**
Senior web content manager   **20×× - Present**
- Increased web traffic 40% in first six months.
- Reduced online marketing costs by 12%.
- Developed and designed company staff portal.

Web consultant   **20×× - 20××**
- Advised and improved traffic for 22 Fortune 500 companies and numerous startup companies.
- Worked with marketing team in search engine optimization (SEO).

Digital A Company   **Tampa, FL**
Web analyst   **20×× - 20××**
- Analyzed online marketing data for new business opprtunities.
- Optimized more than 200 professional websites.

EDUCATION

Information Technology, 20××. University of Colorado, Boulder, Colorado

TECHNICAL SKILLS

Proficient in CMS, Sitefinity, Ektron, XML, and CSS.

單字

- cell 手機（=cellphone）
- senior 資深、有經驗的
- strategy 策略
- collaboration 合作、協力
- cross-media 跨媒體（活用多種媒體行銷技巧）
- expertise 專業
- proactive 主導的、先行的
- web traffic 網路流量（拜訪網站的使用者數，或是流動的數據量）
- company portal 公司內部網路
- numerous 很多的、大量的
- optimization 最佳化
- technical skills 技能

美國百大企業文書撰寫祕訣！

Resume 是應徵工作時提交的文件，也就是「履歷」的意思。有經驗者的履歷內容應以經歷為主，建議減少個人資訊，以數據為基礎盡可能詳述經歷。將招聘公告中提到的關鍵字及公司要求的條件、技術、經歷融入履歷內容中，可以為求職加分。初階求職者的履歷則相對著重學歷。以第三人稱寫作是基本原則，近年來常見的編排順序為聯絡方式、概要、核心要點、經歷、學歷、其他。

實務負責人的一句話！

使用易讀的排版，盡量將篇幅濃縮在一頁之內。寫完後要詳細確認錯字、字體等。可加入一點色彩以免看起來太過平凡，或是放上領英（LinkedIn）連結，使履歷更突出。

文書架構詳細解析

Mark C. Clark

❶ 2220 Conner St. Gulfport, FL 39501 | Cell: (494) 7834-5684 | markclark@msn.com

WEB CONTENT MANAGER

❷ Senior content manager with 15 years of online marketing experience. Managed more than 150 websites for various companies.

❸
- Growth strategies
- Multimedia experience
- Proactive e-commerce
- Superior collaboration
- Cross-media expertise
- Customer service oriented

PROFESSIONAL EXPERIENCE

Webgem Corporation — Jacksonville, FL
Senior web content manager — 20XX - Present
- Increased web traffic 40% in first six months.
- Reduced online marketing costs by 12%.
- Developed and designed company staff portal.

❹ **Web consultant** — 20XX - 20XX
- Advised and improved traffic for 22 Fortune 500 companies and numerous startup companies.
- Worked with marketing team in search engine optimization (SEO).

Digital A Company — Tampa, FL
Web analyst — 20XX - 20XX
- Analyzed online marketing data for new business opprtunities.
- Optimized more than 200 professional websites.

❺ EDUCATION

Information Technology, 20XX. University of Colorado, Boulder, Colorado

❻ TECHNICAL SKILLS

Proficient in CMS, Sitefinity, Ektron, XML, and CSS.

---

❶ 聯絡方式（Contact Information）
寫出姓名、居住地、電話（手機）號碼、電子信箱等。近年來也會放上領英（LinkedIn）網站的個人首頁連結。

❷ 概述（Summary）
寫出欲應徵的職務、本人的經歷及特長。

❸ 核心要點（Highlight）
這部分也可命名為 Qualities 或 Areas of Expertise，用來描述自身的優點和特長。

❹ 經歷（Experience）
詳細記載曾經任職的公司名稱、工作地點、職責、工作期間、實際成就。

❺ 學歷（Education）
詳細記載學校名稱、學校所在地、主修科系、畢業年度。當經歷稍嫌不足的時候，最好提出校內活動、修習課程、是否為優秀畢業生等具體細節。

❻ 其他（Other）
記載本人具備的技能和持有的證照、獲獎經歷、社會活動等。最好不要提到與職務無關的事項，或駕照之類等常見證照。

撰寫文書所需的主要用語

01 **with... year(s) of... experience**　具有⋯年⋯的經驗的
Senior accountant **with** 10 **years of** fund managing **experience**.
具有 10 年基金管理經驗的資深會計師。

02 **increase web traffic/revenue by...**　使網站流量 / 營收增加至⋯
Increased revenue by 30% for all company products.
公司全部產品營收增加 30%。

03 **reduce costs by...**　使花費減少至⋯
Reduced marketing **costs by** over 20% in one quarter.
在一季內使行銷成本減少 20% 以上。

04 **graduate from...**　畢業於⋯
Graduated from Seohan University with a degree in physics.
畢業於瑞韓大學物理系。

05 **build rapport with...**　與⋯建立融洽的關係
Built rapport with local clients for future business opportunities.
為追求未來的商機,與地方客戶建立融洽的關係。

06 **(be) responsible for**　負責⋯
Responsible for writing reports and correspondence for senior management.
負責為高層管理人員撰寫報告及信件。

07 **meet/achieve/exceed sales goals**　達到 / 達成 / 超過銷售目標
Met sales goals for first half of 20××.
達到 20×× 年度上半年的銷售目標。

08 **proficient in...**　精通⋯
Proficient in Excel and creating spreadsheets.
精通 Excel 及製表。

小測驗

01. (　　　　　　　　　) writing expense reports.
　　負責製作費用報表。

02. (　　　　　　　　　) by 55,000 dollars at Terra Company.
　　泰拉公司將成本減少 55,000 元。

03. (　　　　　　　　　) XML and HTML.
　　精通 XML 和 HTML。

04. (　　　　　　　　　) with Canadian customers.
　　與加拿大客戶們建立融洽的關係。

DANIEL ALMEIDA

daniel@maruchs.edu | C: (755) 7834-5684 | LINKEDIN.com/in/daniel-almeida

Summary

Ambitious and results-driven individual with strong academic credentials and substantial intern experience. Independent professional with high energy and superb communication skills.

Highlights

- Attention to detail
- People oriented
- Multitasker
- Good communicator

Experience

Rogu Ad   **Chicago, IL**

Ad Sales Intern   **Summer 20×✕**

- Worked closely with sales team on organizing strategic plans.
- Coordinated ad campaigns for both radio and print.

MGT Center   **Springfield, IL**

Sales Intern   **Summer 20×✕**

- Interacted with clients in all sales stages.
- Made outbound telephone calls to prospective clients (50 calls a day).
- Arranged orders and deliveries for more than 50 clients.

Education

Maruchs College   **Chicago, Illinois**   **20×✕-20×✕**

Marketing   Magna Cum Laude

Activities

- Vice President, Student Bureaucracy Club, 20×✕
- Treasurer, Beta Gamma Honors Society, 20×✕-20-×✕

單字

- ambitious 有野心的
- results-driven 以結果為導向、追求實績的
- academic credentials 學歷、學位
- substantial 很多的
- independent 獨立的
- superb 優秀的
- attention to detail 注重細節
- multitasker 多重任務處理者、多工處理者
- coordinate 協調、聯絡以訂定某事
- interact 互動
- outbound call 致電外部
- arrange 安排、準備

33

Diana Gardner

Virtual Sales Account Manager at Utley
Utley • Indiana State University
Evansville, Indiana • 500+

| Send inMail | Connect |

I graduated from the Scott College of Business at Indiana State University with a degree in business administration. I recently joined Utley Systems through the CRAD program and I'm currently working with and learning account management.

Experience

Virtual Sales Account Manager

Utley Systems
Dec 20✕✕– Present • 5 mos
Evansville, Indiana

As a Virtual Sales Account Manager in Evansville I drive business in the entire Midwestern region. My responsibilities include:
• Cultivate client relationships and consult with decision makers of small businesses and Fortune 500 companies to achieve their business objectives
• Sales forecasting on a weekly basis

Associate Sales Representative

Kay Fusion Inc.
Jan 20✕✕ – Dec 20✕✕ • 2 yrs
Bloomington, Indiana
• Built rapport with potential clients online and through personal visits
• Responsible for timely billing and processing of 75% of company revenue
• Exceeded yearly sales goals for both 20✕✕ and 20✕✕

Education

Indiana State University

B.S. in Business Administration, Scott College of Business
20✕✕-20✕✕
Dean's List 20✕✕-20✕✕

戴安娜・加德納
虛擬銷售帳號管理人
厄特利・印第安納州立大學
印第安納艾凡士維 ・500 人以上（人脈）
發送訊息 / 建立關係

我畢業於印第安納州立大學斯考特商學院，並得到商業管理學位。最近我透過 CRAD 計畫加入厄特利系統，目前正在經營並學習帳號管理。

經歷
虛擬銷售帳號管理人
厄特利系統
20×× 年 12 月 – 現在（5 個月）
印第安納艾凡士維
身為一個在艾凡士維工作的虛擬銷售帳號管理人，我負責推動中西部全區的業務。我負責的業務包括：
· 建立客戶關係並與中小企業及財星全球 500 強企業的決策者商談，以達成其事業目標。
· 預測每週銷售成績

業務代表
凱富爾斯公司
20×× 年 1 月 – 20×× 年 12 月（2 年）
印第安納布魯明頓
· 透過網路及親身拜訪，與潛在客戶建立融洽關係
· 負責公司 75% 的收入並及時開立帳單
· 20×× 年及 20×× 年皆超過年度銷售目標

學歷
印第安納州立大學
斯考特商學院工商管理學士
20××-20××
院長榮譽榜 20××-20××

- virtual sales 虛擬銷售
- business administration 企業管理學、商學
- recently 最近
- account management 帳號管理
- drive business 推動事業
- business objective 事業目標
- sales forecasting 銷售預測
- weekly basis 每週的
- rapport 融洽關係
- timely 及時的
- Dean's list 院長榮譽榜（優秀學生名單）

求職信

 文書範本 1 有經驗者求職信

▶ 翻譯請參閱 p.202

Gregory Jenkins

3326 Chatham Way, College Park, MD 20741
Cell: (522) 3737-4576
mjenkins@mail.com

Dear Mr. Ken Hillman,

I read with great interest your posting at AngelList for a new network engineer and would like to apply for the job as I'm eagerly pursuing a career in computer networks.

I found Geodirect's model of network streamlining innovative and intriguing. In particular, I have great interest in your firm's use of the latest hardware and security technology in building network infrastructures.

I am currently working as a network technician in the D.C. area and have experience troubleshooting and resolving network issues. One of my areas of expertise is problem solving and would like to continue in this line of work in the future. I believe this skill would be a great asset to your company.

I've attached my resume for your review. Please contact me if you have any questions. I look forward to hearing from you.

Sincerely,

Gregory Jenkins

單字

- **posting** 貼文、公告
- **eagerly** 渴望地、熱切地
- **pursue** 追求
- **streamlining** 精簡化、使（某事）效率提高
- **innovative** 創新的
- **intriguing** 非常有趣的
- **security technology** 安全技術
- **infrastructure** 基礎建設
- **troubleshoot** 排除故障、解決難題
- **resolve** 解決
- **line of work** 職業的種類、工作領域
- **asset** 資產、強項、人才

美國百大企業文書撰寫祕訣 ！

Cover letter 是求職時以信件形式撰寫，隨履歷一同呈交的文書。基本目的是向人事負責人自我介紹，更大的用意在於展現自己是最適合該職缺的人選。雖然在大部分的人事負責人眼中，求職信的重要性不如履歷，但也不乏詳細檢視的負責人。寫求職信的時候，比起複述履歷已提過的內容，最好以補全履歷中未能展現的特點為寫作方向。

實務負責人的一句話 ！

求職信不需要用生硬的語氣，可以盡量展現自己的強項及對該職缺的熱忱。建議用生動的方式，與職務有關的活動和經歷為中心鋪寫，盡可能激起人事負責人的興趣。

文書架構詳細解析

❶

Gregory Jenkins
3326 Chatham Way, College Park, MD 20741
Cell: (522) 3737-4576
mjenkins@mail.com

❷ Dear Mr. Ken Hillman,

❸ I read with great interest your posting at AngelList for a new network engineer and would like to apply for the job as I'm eagerly pursuing a career in computer networks.

❹ I found Geodirect's model of network streamlining innovative and intriguing. In particular, I have great interest in your firm's use of the latest hardware and security technology in building network infrastructures.

I am currently working as a network technician in the D.C. area and have experience troubleshooting and resolving network issues. One of my areas of expertise is problem solving and would like to continue in this line of work in the future. I believe this skill would be a great asset to your company.

❺ I've attached my resume for your review. Please contact me if you have any questions. I look forward to hearing from you.

❻ Sincerely,

Gregory Jenkins

---

❶ 聯絡方式（Contact Information）
和履歷書相同，寫出姓名、居住地、電話號碼、電子信箱等。

❷ 收件人（Recipient）
以 Dear Mr./Ms. 作為開頭，後面接負責窗口的名字。若不知道對方的名字，可以寫 Dear Hiring Manager（招聘負責人）。附帶一提，To Whom It May Concern 和 Dear Sir or Madam 屬於舊式用法，現在不太常用。

❸ 引言（Opening）
用足以引起興趣的句子作為開場白，簡單自我介紹並說明應徵動機。

❹ 本文（Main Body）
在本文中說明自己能夠為公司做出的貢獻，以及自己為何適合該職缺。建議要說出自己對這間公司的了解和相關經歷。

❺ 結論（Conclusion）
傳達對這份工作的渴望，對讀信的負責人致謝，並表示會靜候佳音。

❻ 結尾（Closing）
除了最常見的 Sincerely, 之外，也可以使用 Regards,、Yours truly, 等作結尾。

撰寫文書所需的主要用語

01 **be excited/thrilled to apply** 很高興可以應徵
I **am thrilled to apply** for your firm as a health care technician.
很高興應徵貴公司的醫療技師。

02 **be currently working as...** 目前的工作是…
I **am currently working as** a receptionist at Ritzy Hotel.
我目前的工作是麗思飯店的櫃檯接待人員。

03 **one of my areas of expertise is...** 我的專長領域之一是…
One of my areas of expertise is designing professional banners.
商用橫幅設計是我的專長領域之一。

04 **continue (in) this line of work** 持續從事這個（領域的）工作
I would to **continue in this line of work** in the near future.
近期內我會繼續從事這個領域的工作。

05 **as a recent graduate of...** 身為剛從…畢業的
As a recent graduate of Brown University, I know the area very well.
身為剛從布朗大學畢業的畢業生，我很了解這個地區。

06 **one of my responsibilities is...** 我負責的工作之一是…
One of my responsibilities was selling ads to local businesses.
我負責的工作之一是賣廣告給地方企業。

07 **I learned how to...** 學到…的方法
I learned how to use Office Suite at my previous company.
我在前公司中學到如何活用 Office Suite。

08 **be a great asset to...** 成為…的重要資產 / 強項 / 人才。
I will work feverishly to **be a great asset to** your company.
我會在工作崗位上努力，成為貴公司不可或缺的人才。

小測驗

01. I am (   ) as a botanist in Palm Springs.
我目前的工作是在棕櫚泉當植物學家。

02. My expertise in marketing should be a (   ) to your company.
我的行銷專業將會是貴公司不可或缺的資產。

03. I (   ) to apply for Health International.
很高興有機會可以應徵健康國際公司。

04. I (   ) communicate with people at an early age.
在我小的時候學會如何與人溝通。

Jung-hwan Kim

2F, 397 Sungam-ro, Mapo-gu, Seoul, 03930
Cell: 010-3737-4576
junghwan173@mail.co.kr

Dear Hiring Manager,

My name is Jung-hwan Kim and I'm excited to apply for a position as an entry-level journalist at *Financial Times*.

As a recent graduate of Gangnam University I worked as the editor-in-chief of my school newspaper. One of my many responsibilities was ad sales which opened my eyes to margin and profit. I learned how to create, manage, and maintain professional relationships with business owners around our campus.

I really enjoy reading the advice section on *Financial Times*. It's got a great blend of modern tips, many of which I've taken to heart. I've already thought of my own tips that I could share if I were to contribute to the section.

Please review my attached resume for additional details regarding my activities and achievements. Thank you for your time and consideration.

Best regards,

Jung-hwan Kim

單字

- excited 興奮的、高興的
- graduate 畢業生
- editor-in-chief 總編輯
- ad sales 廣告業務
- open one's eyes 啟發
- margin 利潤、差額
- profit 利益、收益
- business owner 商家
- blend 融合
- take to one's heart 銘記在心
- contribute 做出貢獻
- additional 附加的
- achievement 業績、成就
- consideration 考慮、考量

Unit 06 推薦信

Dear Mr. Lee,

It's my absolute pleasure to recommend Alice Curtis for the assistant HR manager position with your firm. Alice and I worked together at The Terra Company, where I was her manager and direct supervisor from 20×× to 20××.

I thoroughly enjoyed working with Alice in the HR department due to her dependability and hard-working attitude. She has great people skills and has an eye for spotting talented workers. Her educational background was a huge advantage to us so she was responsible for staff training. She developed an effective in-house training program that we still use to raise morale of our employees. She's a true team player and has always inspired those around her.

I highly recommend Ms. Curtis to your team. I'm positive that she will be a valuable addition to your company. Please feel free to contact me at 555-153-4557 or by email if you have any future questions regarding her qualifications and experience.

Best wishes,

Calvin Kuhlman
Director of Human Resources
The Terra Company

單字

- absolute 十足的、絕對的
- pleasure 愉快
- direct supervisor 直屬上司
- thoroughly 徹底地、完全地
- enjoy 樂於
- dependability 可靠
- people skills 人際關係處理能力
- huge advantage 巨大的利益、好處
- staff training 員工訓練
- in-house 公司內部
- morale 士氣
- team player 善於合作者
- inspire 激勵
- valuable addition 寶貴的擴充資源

美國百大企業文書撰寫祕訣！

Recommendation letter 又稱 reference letter。當某人參與公司招聘、申請入學、申請獎學金時，會請他人撰寫推薦信，對其特長、才能、技術等予以評論並表明推薦之意，以促成想要的結果。推薦信的撰寫人通常是長期和推薦對象共事的前職場上司、教授或同事等。要求應徵者提供推薦信的公司不多，多為大學、研究所要求申請入學者準備推薦信。因為推薦信內容講求客觀，撰寫時最好提供可以佐證推薦事項的細節。

實務負責人的一句話！

如果想寫出內容充實的推薦信，應當要求應徵者提供推薦信草稿、履歷、招聘公告等資料。此外，撰寫時最好以自己和應徵者的共同經歷或故事為出發點，以證明自己足夠了解應徵者。

文書架構詳細解析

❶ Dear Mr. Lee,

❷ It's my absolute pleasure to recommend Alice Curtis for the assistant HR manager position with your firm. Alice and I worked together at The Terra Company, where I was her manager and direct supervisor from 20×× to 20××.

❸ I thoroughly enjoyed working with Alice in the HR department due to her dependability and hard-working attitude. She has great people skills and has an eye for spotting talented workers. Her educational background was a huge advantage to us so she was responsible for staff training. She developed an effective in-house training program that we still use to raise morale of our employees. She's a true team player and has always inspired those around her.

❹ I highly recommend Ms. Curtis to your team. I'm positive that she will be a valuable addition to your company. Please feel free to contact me at 555-153-4557 or by email if you have any future questions regarding her qualifications and experience.

Best wishes,

❺ Calvin Kuhlman
Director of Human Resources
The Terra Company

❶ 收件人（Recipient）
以 Dear Mr./Ms. 為開頭，後面接收件人的名字。

❷ 引言（Opening）
告訴對方自己是如何認識應徵者，與應徵者共事多久，並說明自己和應徵者的關係，是否曾親自管理應徵者等。

❸ 本文（Main Body）
本文中描述推薦人的優點、專長、資質及技術等，說明自己推薦應徵者的理由，並盡可能提出具體事例以茲證明。

❹ 結論（Conclusion）
再次強調並積極推薦應徵者，以相信應徵者必將成為公司的寶貴人才作結。此處也可以把自己的聯絡方式提供給人事負責人。

❺ 結尾（Closing）
寫上 Best wishes, 等問候語，署名並寫出職稱、所屬單位。

撰寫文書所需的主要用語

01 **It's my pleasure to recommend...** 很高興能夠推薦…
It's my pleasure to recommend Josh Lim to your organization.
很高興能夠向貴公司推薦喬希·林。

02 **enjoy working with...** 和…共事很愉快
I **enjoyed working with** Ms. Kime during my time at Intel.
我在英特爾工作期間，和金女士留下了愉快的共事經驗。

03 **have great people skills** 善於處理人際關係
John surely **has great people skills**.
約翰有著出色的人際交往能力。

04 **has an eye for...** 對…有眼光、對…有判斷力
She **has an eye for** finding good employees.
她很有為公司尋找人才的眼光。

05 **have known... for... years** 和…認識…年
I **have known** Ms. Ewing **for** two **years**.
我和尤茵女士相識 2 年。

06 **go the extra mile** 特別用心（努力）
Jake always **goes the extra mile** to please his boss.
傑克一向努力滿足上司的需求。

07 **build trust with...** 獲得…的信賴
He strives to **build trust with** contractors and clients.
他致力於與承包商和客戶建立信任感。

08 **be eager to learn...** 渴望學習…
Minhee **is eager to learn** about the poultry business.
敏希渴望學習家禽業的相關知識。

小測驗

01. Vera always (   ) to finish a project on time.
薇拉努力準時完成計畫。

02. I enjoyed (   ) Sam during his time here.
我和山姆留下了愉快的共事經驗。

03. It's my pleasure (   ) Corey Choi.
很高興能夠推薦科里·蔡。

04. Ms. Jang (   ) for coming up with solutions to problems.
張小姐對尋求問題解決方案很有判斷力。

Trinity University, College of Business
Admissions - Executive MBA Program
4495 Glendale Avenue
Pomona, CA 91766

Dear Admissions Committee,

I'm writing to recommend Darius Choi for admission into Trinity University's Executive MBA program. I have known Darius for more than six years when he and I worked together as colleagues at the UP's Pacific Branch here in Seoul.

He was a unique worker in that he always went the extra mile to build trust with clients. As a result, he oversaw 110% sales growth in his last three years here. Prior to that, he helped organize our accounting department which saved the company more than $22,000 in annual operating expenses.

During the years I've known him as a colleague, he is always energetic, optimistic, and eager to learn. It is my belief these qualities would make him an outstanding student for your program. Please feel free to contact me at any point should you have questions about this recommendation.

Best regards,

Seo-young Hwang
Business Plan Manager
UP, Pacific Branch (Seoul, Korea)

單字

- admission 入學
- colleague 同事
- branch 分公司
- unique 獨特的
- trust 信賴、信任感
- oversee 監督
- annual 年度的
- operating expenses 營業成本、營運成本
- energetic 精力充沛的、充滿活力的
- optimistic 樂觀的
- eager to... 對…充滿渴望
- outstanding 出眾的、傑出的

PART 3
Basic Work Documents

基本工作文書

ESG Report

文書範本 1 傳真封面　　　　　　　　　　　　▶翻譯請參閱 p.206

St. Paul Health Services

Geriatric Consult Team
Westland Office
St. Paul, MN 9524
Phone : 070-393-4800
Fax : 070-393-4880

FAX COVER SHEET

www.stpaulhealthservices.org

Date: *June 6, 20××*　　　　　　　Pages: *5*
　　　　　　　　　　　　　　　　(Including cover sheet)

To
Name: *James E. Walden*
Fax: *070-7836-4880*
Phone:

From
Name: *Michael Loza*
Fax: *070-393-4880*
Phone: *070-393-4844*

Message: (Please do <u>NOT</u> include confidential or personally identifiable information on the cover sheet.)

James,

Here's a copy of the signed sales agreement. This is to confirm that we agree to the conditions that you laid out yesterday. Please let me know if you need anything else.

Mike

單字

- geriatric 老年醫學的、老年人的
- consult 顧問、諮商
- fax cover sheet 傳真封面
- include 包含
- confidential 機密
- personally identifiable information 個資
- signed 已簽名的
- agreement 協議、協議書、合約書
- confirm 確定
- agree to... 同意（答應）…
- condition 條件
- lay out... 列出…

美國百大企業文書撰寫祕訣 ！

Fax（傳真）是一種利用電話線路，傳送文件複印本的電子通訊技術。使用頻率隨著電子郵件的問世漸漸式微，不過還是有不少公司仍在使用，尤其經常活用於傳送需要簽名的文件。近來也出現使用網路的 eFax（網路傳真），以電腦取代傳真機，不只便於確認文件內容，還省了電話費。基本上傳真應載明發件人、發件目的、傳真號碼等資訊，所以建議加上 fax cover sheet（傳真封面頁）。

實務負責人的一句話 ！

傳真封面不是必要的，不過最好還是附上包含傳真號碼的聯絡方式，以便收件人回覆。特別是首次透過傳真聯絡時，姓名、電話號碼、傳真號碼都是應附的基本資訊。

文書架構詳細解析

St. Paul Health Services
Geriatric Consult Team   FAX COVER SHEET
Westland Office
❶ St. Paul, MN 9524
Phone : 070-393-4800
Fax : 070-393-4880

www.stpaulhealthservices.org

---

❷ Date: *June 6, 20××*   Pages: 5
                                 (Including cover sheet)

To	From
Name: *James E. Walden*	Name: *Michael Loza*
Fax: *070-7836-4880*	Fax: *070-393-4880*
Phone:	Phone: *070-393-4844*

❸

Message: (Please do <u>NOT</u> include confidential or personally identifiable information on the cover sheet.)

James,

❹ *Here's a copy of the signed sales agreement. This is to confirm that we agree to the conditions that you laid out yesterday. Please let me know if you need anything else.*

Mike

---

❶ **公司資訊（Company Information）**
記載發件人的公司名稱、地址、電話號碼、傳真號碼等資訊。

❷ **日期與頁數（Date & Pages）**
記載發送傳真的日期，及包含傳真封面在內的完整頁數。

❸ **收件人及發件人（To & From）**
留下收件人和發件人的姓名及聯絡方式（電話號碼、傳真號碼）。務必寫明發件人的傳真號碼，以便收件人回覆。

❹ **訊息內容（Message）**
表明傳真文件的目的。也可在此處提出要求，請收件人針對傳送的文件做出相關的措施或回應。

01 **here's a copy of...** 這是…的複印本
Here's a copy of the receipt that you requested.
這是您索取的發票影本。

02 **let me know if...** 如果…的話，請與我聯絡
Let me know if you have any questions.
如有任何疑問，請與我聯絡。

03 **serves as a(n)...** …的功能是…
This letter **serves as a** response to your latest request.
這封信是為了回應您最近提出的要求。

04 **inquiry/letter/request dated...** 於（日期）提出的詢問／信件／要求
This is to reply to your **letter dated** May 15, 20××.
這是為了回覆您於 20×× 年 5 月 15 日的來信。

05 **if you need anything else** 如果有其他需求
Please contact me **if you need anything else**.
如果您有其他需求，請與我聯絡。

06 **be of any further assistance** 進一步協助
Please let me know if I can **be of any further assistance**.
如果需要我提供進一步的協助，請與我聯絡。

07 **this is to confirm...** 這（封信）是為了確認…
This is to confirm that the price in the quote is our best offer.
這封信是為了向您確認，報價單的價格已經是我們最佳的價格。

08 **lay out conditions** 提出（列出）條件
The **conditions are laid out** in the last page.
條件列於最後一頁。

小測驗

01. (   ) of the invoice.
這是發票影本。

02. This (   ) our acceptance of the shipping conditions.
這封信是為了向您確認我們接受的運送條件。

03. (   ) if you have any inquiries.
如有任何疑問，請與我聯絡。

04. This is a reply to your (   ) March 30.
這是為了回覆您於 3 月 30 日的來信

Centralnet Mutual

4210 Emma Street
Clovis, TX 88101
Telephone: (350) 388-6931
Fax: (350) 388-6930
www.centralnetmutual.com

October 13, 20××

Kathleen Long
Sentinnel News
3757 Ocala Street
Orlando, FL 32805

Dear Ms. Long:

This letter serves as a response to your inquiry dated October 10, 20×× on Centralnet Mutual's holding of certain documents.

1. Paper copy of electronics waste disposal and recycling contracts for 20××, 20×× and 20××.

   • **Centralnet Mutual no longer holds copies of such contracts for those years.**

2. Paper copy of hard drive disposal contracts for 20××, 20×× and 20××.

   • **Centralnet Mutual no longer holds copies of such contracts for those years.**

Please let me know if I can be of any further assistance.

Best regards,

Wallace Amos

Wallace V. Amos
Director, Audit Office
Centralnet Mutual

單字

- serve as... 有…的作用
- response 回覆、回信
- inquiry 詢問
- dated... …（日期）的
- hold 持有、保管
- document 文件、檔案

- contract 合約（書）
- no longer... 再也不…
- waste disposal 廢棄物處理
- recycling 回收
- hard drive 硬碟
- further assistance 進一步協助

報告書

COMPANY DRESS CODE ANALYSIS
Lynn M. Lowe - Human Resources Department
April 11, 20××

Table of Contents
Objective
Current Status
Analysis
Recommendation

Objective
The objective of this report is to report the findings of our review of the company dress code and make recommendations based on those findings.

Current Status
The company currently requires that all employees adhere to a strict formal dress code. This is to establish a professional working environment made necessary by the frequency of visits from clients, contractors and partners. Permissible attire includes suits, sports jackets, pants and skirt suits that are appropriate to a formal business setting.

Analysis
Outside of the Sales Department there isn't a great frequency of meetings with outside clients and partners. Workers seem to be genuinely uncomfortable with formal dress and have an assertion that a more casual code would raise worker morale and foster more freedom and creativity in the workplace. An anonymous survey conducted on April 2, 20×× reveals that a vast majority of employees, 83%, prefer a more casual dress code in the workplace.

Recommendation
The HR Department recommends that the dress code for the company be revised from formal dress to business casual for all employees. Business casual refers to suits, pants, jackets, shirts, skirts and dresses that, while not formal, are appropriate for a business environment. We trust that employees will use good judgment in determining what's appropriate and not appropriate for the workplace. Workers who wear attire that is deemed inappropriate or unprofessional will be dealt with on an individual basis.

單字

- objective 目的
- analysis 分析、評估
- permissible 許可的
- appropriate 適合的
- frequency 頻率
- assertion 斷言
- judgment 判斷、判斷力
- determine 決定

美國百大企業文書撰寫祕訣！

Report（報告書）的目的是為了報告特定業務狀況、執行項目或是研究及檢討結果。報告書分成很多種，包括 informational report（資訊報告）、expense report（開銷報告）、progress report（進度報告）、appraisal report（評估報告）、recommendation report（建議報告）、sales report（行銷報告）、project report（專案報告）、handover report（交接報告）等。報告書可能會要求報告對象回覆，但非必要，每種報告書的目的也各有不同。不過不管是哪一種報告書，目的都是為了將資訊傳達給對方，因此寫作的重點在於把內容梳理清楚。

實務負責人的一句話！

撰寫工作上的報告書時，比起按照一定的格式撰寫，更重要的是正確且具體地寫出對方須知的內容，及寫清楚欲傳達的內容摘要。

文書架構詳細解析

❶ COMPANY DRESS CODE ANALYSIS
Lynn M. Lowe - Human Resources Department
April 11, 20××

❷ Table of Contents
Objective
Current Status
Analysis
Recommendation

Objective
The objective of this report is to report the findings of our review of the company dress code and make recommendations based on those findings.

Current Status
❸ The company currently requires that all employees adhere to a strict formal dress code. This is to establish a professional working environment made necessary by the frequency of visits from clients, contractors and partners. Permissible attire includes suits, sports jackets, pants and skirt suits that are appropriate to a formal business setting.

Analysis
Outside of the Sales Department there isn't a great frequency of meetings with outside clients and partners. Workers seem to be genuinely uncomfortable with formal dress and have an assertion that a more casual code would raise worker morale and foster more freedom and creativity in the workplace. An anonymous survey conducted on April 2, 20×× reveals that a vast majority of employees, 83%, prefer a more casual dress code in the workplace.

Recommendation
❹ The HR Department recommends that the dress code for the company be revised from formal dress to business casual for all employees. Business casual refers to suits, pants, jackets, shirts, skirts and dresses that, while not formal, are appropriate for a business environment. We trust that employees will use good judgment in determining what's appropriate and not appropriate for the workplace.

❶ 標題（Heading）
載明報告書的題目、撰寫者、日期等資訊。

❷ 目次（Table of Contents）
寫出報告書的目次。

❸ 本文（Main Body）
確定本文架構後再寫出各部分的內容。本文架構依報告書種類有各式各樣的型態，既定的報告書本文以「目的→現況→分析」的順序撰寫。

❹ 結論（Conclusion）/ 建議事項（Recommendation）
對前述的內容做出最終結論。如為調查或研究報告書，只要記載調查及研究結果即可。也會根據結論，針對報告書中探討的問題及案例提出建議事項。

01 **the objective of this report is to...** 這份報告的目的是⋯
The objective of this report is to provide a current overview of the situation.
這份報告的目的是概述當前情況。

02 **make recommendations based on...** 基於⋯提出建議
We would like to **make recommendations based on** the results of the lab test.
我們將根據實驗結果提出一些建議。

03 **(have) an assertion that...** 提出⋯的主張
Comcast **had an assertion that** the press went too far.
康卡斯特公司提出媒體太超過的主張。

04 **be appropriate for...** 適合⋯
This extreme measure **was** not **appropriate for** this plant.
這種極端手段不適用於這間工廠。

05 **occur/last from... to...** 發生在⋯到⋯之間 / 從⋯持續至⋯
The sale **lasts from** August 10 **to** 31.
折扣將從 8 月 10 日持續至 31 日。

06 **recommend that...** 建議⋯
The committee **recommends that** the dress code be modified.
委員會建議調整服裝規定。

07 **be well executed** 妥善執行
Walmart's marketing strategy **was well executed** on many levels.
沃爾瑪的行銷策略在許多層面都有妥善執行。

08 **be dealt with** 被處理、被處置
Rule breakers will **be dealt with** severely.
違規者將受到嚴格處置。

小測驗

01. The conference (   ) June 30 to July 1.
會議將從 6 月 30 日持續至 7 月 1 日。

02. The objective of this (   ) provide a possible plan of action.
這份報告的目的是提供可行的對策方案。

03. Employees (   ) benefits could be better.
職員們主張福利可以再更好。

04. The crisis must be (   ) immediately.
應該立刻處理這個危機。

After Action Review
Department of Environmental Protection
U.S. Oil and Refining
October 21, 20✕✕

This report provides an After Action Review (AAR) of the Tacoma refinery fire incident and makes follow-up recommendations. The incident occurred from February 13, 20✕✕ to February 16, 20✕✕ at the Tacoma refinery in Tacoma, Washington.

This AAR contains four sections:
1. Summary of the Incident
2. Cause of the Incident
3. Effectiveness of the Response
4. Recommendations to Prevent Future Incidents

1. Summary of the Incident
On Tuesday, February 13, 20✕✕ at approximately 0850, leaking of natural gas was discovered at the 9D tank. A fire soon broke out after about 20 minutes at 0910. At 0934, Gas Fire Control (GFS) was contacted for help. GFS staff began to arrive at the incident site at roughly 1100. Due to the continuous leak of gas, the fire was only brought under control by 1300 on February 16.

2. Cause of the Incident
After a thorough investigation, the GFS investigative committee made the conclusion that the incident was caused by too much pressure in the 9D tank which allowed large quantities of methane gas to be emitted thus igniting the fire.

3. Effectiveness of the Response
GFS's response was well executed. Initial arrival was within two hours of the incident. Once they arrived, responders secured the site and addressed immediate safety concerns. However, there were not enough personnel at the emergency site at least in the first few critical hours to properly combat the situation.

4. Recommendations to Prevent Future Incidents
a. The Department of Environmental Protection should develop a separate response protocol for major fire incidents. The protocol should identify and address proper emergency standards and measures.
b. A regular check-up should be started and maintained by the quality staff to focus on gas leaks particularly pertaining to that of methane gas.

單字

- refinery 煉油廠
- follow-up recommendation 後續建議
- effectiveness 效果、效率
- approximately 大略、約
- bring under control 抑制、鎮壓
- emit 散發、發射
- ignite 點燃、使發光
- address 審慎處理

會議記錄

文書範本 1　每月會議記錄　　▶翻譯請參閱 p.210

Meeting Minutes

Planning Department
Monthly Team Meeting
September 13, 20××

Meeting called to order at 4:30 P.M. by meeting chair Krista Scott.

Members present

Krista Scott (chair)
Sheryl Walker
Mike Kiefer

Members not present

(none)

Approval of Minutes

Minutes from August 16, 20×× meeting approved without modification.

Motion

Motion: Chair Krista Scott made a motion to hold product testing on October 1, 20××.
Vote: 3 for, 0 opposed
Resolved: Motion carried.

Proceedings

• Monthly finance report provided by chair Krista Scott.
• Online promotion update provided by Sheryl Walker.
• Mike Kiefer announced that he had recently hired a new secretary, Georgia Denham.

Closing

October 15, 20××, 4:30 P.M. was designated as the date and time of the next meeting.
Meeting adjourned at 6:00 P.M. by meeting chair Krista Scott.

單字

- meeting chair 主席
- call to order 開會
- approval 批准
- motion 動議
- resolved 決議
- carried 通過（議案等）
- proceedings 進行過程
- provide 提供、提出、報告
- announce 宣佈
- closing 結尾
- designate 指定
- adjourn 使中止、結束

美國百大企業文書撰寫祕訣！

Minutes（會議記錄）是記錄商務會議的討論內容，並留存保管的文書。從開會時間到決議事項都應詳細記錄下來，讓未出席者也能迅速掌握會議內容。會議記錄的功能是提供給出席者和相關人員，向上司報告或參考的會議內容，需要系統性的管理。

實務負責人的一句話！

會議記錄應排除主觀的摘要或意見，只客觀地記錄會議中提到的事實。會議記錄要能夠證明會議上決定的事項，因此撰寫重點是準確且毫無遺漏地記錄內容。

文書架構詳細解析

Meeting Minutes

❶ **Planning Department**
Monthly Team Meeting
September 13, 20××

Meeting called to order at 4:30 P.M. by meeting chair Krista Scott.

Members present
Krista Scott (chair)
❷ Sheryl Walker
Mike Kiefer

Members not present
(none)

Approval of Minutes
❸ Minutes from August 16, 20×× meeting approved without modification.

Motion
❹ Motion: Chair Krista Scott made a motion to hold product testing on October 1, 20××.
Vote: 3 for, 0 opposed
Resolved: Motion carried.

Proceedings
• Monthly finance report provided by chair Krista Scott.
❺ • Online promotion update provided by Sheryl Walker.
• Mike Kiefer announced that he had recently hired a new secretary, Georgia Denham.

Closing
❻ October 15, 20××, 4:30 P.M. was designated as the date and time of the next meeting.
Meeting adjourned at 6:00 P.M. by meeting chair Krista Scott.

---

❶ 標題（Heading）
記錄會議討論的題目、開會日期及開始時間等。

❷ 出席／缺席名單（Attendees/Absentees List）
完整記錄出席者和缺席者的姓名。有必要時應連同出席者和缺席者的職位一併記載。

❸ 核可會議記錄（Approval of Minutes）
出席者確認並同意會議記錄內容無誤。

❹ 動議（Motion）
如有提出動議，則記載動議內容及其結果。

❺ 會議過程（Proceedings）
詳細記載會議中提到的報告及討論等內容。

❻ 結尾（Closing）
記錄決議事項、下次開會的日期及時間、會議結束時間等內容。

撰寫文書所需的主要用語

01 **called to order by... at...** 於…由…宣布開會
Meeting **called to order by** Chair Peterson **at** 5:00 P.M.
於下午 5 點由會議主席彼得森宣布開會。

02 **approved without modification** 核可不需修正
Last meeting's minutes **approved without modification**.
核可上次的會議記錄，不需修正。

03 **made a motion to...** 提出…的動議
Leslie Hwang **made a motion to** introduce the beta version of Secreta 2.0.
雷斯理·黃提出推出斯柯塔 2.0 測試版的動議。

04 **consensus (was) reached that...** 達成…的共識
Consensus was reached that construction of the high rise should begin immediately.
已達成高樓建築應立即開始施工的共識。

05 **be designated as the date and time of...** 被指定為…的日期及時間
November 21, 20××, 11 A.M. **was designated as the date and time of** the next meeting.
指定下次開會的日期及時間為 20×× 年 11 月 21 日上午 11 點。

06 **meeting adjourned at... by...** 會議由…宣佈於…結束
Meeting adjourned at 11:30 A.M. **by** chair Susan Lee.
蘇珊·李主席宣布會議於上午 11:30 結束。

07 **review the matter further** 進一步探討某事
The chair and vice chair agreed to **review the matter further**.
主席和副主席決定進一步探討該議題。

08 **due to no further business** 因為沒有其他事務
Due to no further business, the security meeting was adjourned at 15:50.
因為沒有其他要討論的事務，安全會議於 15:50 結束。

小測驗

01. (   ) to launch the new product.
    達成推出新產品的共識。

02. The plan was (   ).
    這項計劃已被核可，不需修正。

03. August 16, 2 P.M. (   ) as the date and time of the meeting.
    指定開會日期時間為 8 月 16 日下午 2 點。

04. Mr. Davis agreed to (   ) further.
    戴維斯先生決定進一步探討該議題。

Minutes of Marketing Strategy Meeting of March 3, 20✕✕

CALL TO ORDER

Meeting was called to order at 9:35 A.M. by Chair Richard Chung.

ATTENDEES

Chair Richard Chung, Marketing Manager
Rhea Haynes, Sales Manager
John Maya, Assistant Manager, Marketing
Karla Jennings, Junior Manager, Marketing

ABSENTEES

Andres Brown, Assistant Manager, Sales

Motion

Motion: Sales Manager Rhea Haynes made a motion to eliminate 5% discount coupons.
Vote: 1 for, 3 opposed
Resolved: Motion failed.

DISCUSSION

- Briefing of ongoing online marketing campaign by John Maya.
- Alternatives to online advertising discussed such as radio and TV.
- Consensus reached that budget needs to be raised for more aggressive campaigns.

CONCLUSION

Participants agreed to review the matter further before determining definite changes to marketing strategy on the next meeting at March 10, 20✕✕.

ADJOURNMENT

Due to no further business, the meeting was adjourned at 11:15 A.M. by Chair Richard Chung.

Minutes submitted by designated minutes taker, Clara Johnson.

單字

- attendee 出席者
- absentee 缺席者
- eliminate 清除
- failed 否決
- briefing 簡報、報告
- budget 預算
- aggressive 攻擊性的
- agree to... 同意⋯
- determine 決定
- further business 更多事務
- submit 提交
- minutes taker 會議記錄人員

工作交接文件

文書範本 1　生產管理人員交接文件

▶ 翻譯請參閱 p.212

Handover Note

Job title:	Production Supervisor
Date of handover note:	May 13, 20××
Handed over by:	Joseph Johnstone
Taken over by:	Judith Moss

Job description

Supervising the company's beverage production at Dyer plant to meet customer needs and achieve company standards related to cost, waste, safety, productivity and line efficiency.

Key documents

• 20×× operation and production schedule
• Health and safe regulations manual

Key responsibilities

• Develop production schedule for Dyer plant on an annual basis
• Achieve performance standards with regards to quality and safety
• Oversee equipment and its reliability to minimize downtime
• Build personnel capability through employee relations, hiring, training and effective communication

Status of recent and current projects

• Maintenance of production lines number 6 and 7 (Ongoing)
• Review of health and safety regulations (Due May 15, 20××)

單字

• supervisor　主管、監管人
• hand over　移交
• take over　承接
• meet customer need　滿足客戶需求
• achieve company standard　達到公司標準
• line efficiency　生產線效率

• annual basis　年度
• oversee　監督
• reliability　可靠度
• minimize　使…降到最低
• downtime　停機時間
• maintenance　保養

美國百大企業文書撰寫祕訣！

Handover document（交接文件）是前任職員離職或調職時，寫給接任職員的文件，目的是幫助接任職員掌握並理解工作內容。除了寫下職務相關工作、主要問題、目前正在進行的事項等之外，這份文件也是是否妥善交接的證明。有時也會依情況寫出相關人員的聯絡方式、相關檔案位置、報告流程等，所以亦被稱為 Handover note（交接記錄）或 Handover report（交接報告）。

實務負責人的一句話！

交接文件不用寫得過於詳細。重點在於明確寫出前任職員的工作內容，以利接任職員了解。

文書架構詳細解析

Handover Note

	Job title:	Production Supervisor
❶	Date of handover note:	May 13, 20××
	Handed over by:	Joseph Johnstone
	Taken over by:	Judith Moss

Job description

❷ Supervising the company's beverage production at Dyer plant to meet customer needs and achieve company standards related to cost, waste, safety, productivity and line efficiency.

Key documents

❸ • 20×× operation and production schedule
• Health and safe regulations manual

Key responsibilities

❹ • Develop production schedule for Dyer plant on an annual basis
• Achieve performance standards with regards to quality and safety
• Oversee equipment and its reliability to minimize downtime
• Build personnel capability through employee relations, hiring, training and effective communication

Status of recent and current projects

❺ • Maintenance of production lines number 6 and 7 (Ongoing)
• Review of health and safety regulations (Due May 15, 20××)

❶ **標題（Heading）**
記載職稱、撰寫日期、前任職員的姓名、接任職員的姓名等。

❷ **職務說明（Job Description）**
說明前任職員的職務，也就是接任者將負責的工作。

❸ **重要文件（Key Documents）**
寫出該業務有哪些相關文件，也連同記錄文件保管位置。

❹ **主要業務（Key Responsibilities）**
依業務（責任）的重要程度列出業務內容。最好以主要關鍵字為中心撰寫，以利讀者快速掌握重點。

❺ **近期及目前計畫現況（Status of Recent and Current Projects）**
記錄最近結束的業務，和最近正在進行的業務明細。

撰寫文書所需的主要用語

01 **be handed over by...** 由…進行交接
The responsibility **was handed over by** James Phillips.
這份業務由詹姆斯·菲利普斯交接。

02 **be taken over by...** 由…接手
The project **was taken over by** the CS department.
這項計畫由 CS 部門接手。

03 **meet customer needs** 滿足客戶的需求
Meet customer needs based on big data.
根據大數據來滿足客戶的需求。

04 **achieve company standards** 達到公司的標準
Achieve company standards in the production of AK batteries.
AK 電池的生產有達公司的標準。

05 **status of recent and current projects** 近期及目前的計畫現況
This is the **status of recent and current projects**.
這是近期及目前的計畫現況。

06 **ensure that...** 確認（確保）…
Ensure that all rules are being observed.
確保遵照所有規定。

07 **be in accord with...** 和…相符
The report needs to **be in accord with** the other department's documents.
報告應與其他部門的文件一致。

08 **minimize downtime** 將中斷（停機）時間縮到最短
Minimizing downtime is of utmost priority.
首要任務是將停機時間縮到最短。

小測驗

01. Achieve (   ) regarding the development of new vehicles.
    新車的開發有達到公司的標準。

02. You are to (   ) all deadlines are met.
    你必須確認是否能如期完成。

03. The duties are to be (   ) by Jack Mills.
    任務由傑克·米爾斯接手。

04. (   ) is critical during production.
    重要的是在生產時，將停機時間縮到最短。

Handover Report

Written by: Gary Turner
Job Title: Technical writer
Date of Handover: July 11, 20××

Brief description of duties:

- Work closely with engineers and product managers to author comprehensive user and administrator software documentation
- Ensure that all technical manuals are uniform in style and are in accord with actual product functions

Supervisor and reporting procedure:

Mr. Anthony Nunez (Written activity report to be submitted every Friday by 18:00)

Regular meetings, reports or procedures:

- Weekly team meeting on Mondays at 9:00

Key Documents and reference material (attached with this report):

- Internal glossary of terms (Last updated June 20, 20××)

Status of recent and current projects:

- Development of AIPK 2.0 User Manual (Due August 31, 20××)
- Constant updating of internal glossary

Where to find work files (hardcopy and electronic):

"TW Team" folder at company server

Contacts (internal and external)

Name	Title	Phone No.	Email Address
Anthony Nunez	Supervisor	070-383-5844	anunez@mail.com
Jamie Hwang	Team member	070-383-5850	jhwang@mail.com
Jessie Bedford	Team member	070-383-5851	jessieb@mail.com

單字

- duty 職務、工作
- author 撰寫、記述
- comprehensive 綜合的、全面的
- documentation 文件、文件化
- uniform 一律的、一致的
- reference material 參考資料
- status 現況
- glossary 詞彙表
- hardcopy 紙本資料
- electronic 電子的
- internal 內部
- external 外部

文書範本 1　網路釣魚郵件公告　　　▶翻譯請參閱 p.214

Identifying "Phishing" Emails

Recently email messages containing attachments or links to non-Northshore websites have been received by Northshore members. Most often, these attachments are malicious and should not be opened.

If any email or message asks for passwords or credit card information you are advised to ignore it and contact Northshore as soon as possible. It is Northshore's policy to never ask for personal or account information. Be on the lookout for any message asking for any of the following information:

• Northshore account ID and password
• Social security number
• Full credit card number
• Credit card CCV code

If you receive a suspicious email, please notify Northshore Support by forwarding the email to support@northshore.com.

單字

- phishing 網路釣魚
- most often 最常、一般、大部分
- malicious 惡意的
- credit card 信用卡
- be advised 勸告、通知
- policy 政策、方針

- ask for... 要求⋯
- personal information 個人資訊
- account information 帳號資訊
- social security number 社會安全碼
- suspicious 可疑的
- forward 轉寄

美國百大企業文書撰寫祕訣 !

Notice（公告）是用來正式告知資訊或警告的文書。不只可以透過書面、電子郵件、網站，口頭傳達也是公告的形式之一。公告的內容就跟它的形式一樣多變，例如安全、會議、召回、活動、終止服務等訊息都可以經由公告傳達。大部分的公告都包含重要資訊，因此必須以核心主旨，撰寫正確的內容。

實務負責人的一句話 !

公告是會呈現在許多人眼前的文書，所以寫作時不僅要謹慎，也要遵照事實。此外，內容中務必包含公告原因（目的），且應明確描述，避免含混不清。

文書架構詳細解析

---

❶ Identifying "Phishing" Emails

❷ Recently email messages containing attachments or links to non-Northshore websites have been received by Northshore members. Most often, these attachments are malicious and should not be opened.

❸ If any email or message asks for passwords or credit card information you are advised to ignore it and contact Northshore as soon as possible. It is Northshore's policy to never ask for personal or account information. Be on the lookout for any message asking for any of the following information:

- Northshore account ID and password
- Social security number
- Full credit card number
- Credit card CCV code

❹ If you receive a suspicious email, please notify Northshore Support by forwarding the email to support@northshore.com.

---

❶ 標題（Title）
為了讓讀者容易閱讀，應採取醒目的標題。例如加大字型或是使用粗體、彩色來突出標題。

❷ 引言（Introduction）
說明公告目的或是發布公告的原因等背景資訊。

❸ 本文（Main Body）
本文開始提出公告的主要內容。例如讀公告的人應注意的事項，應採取的措施，應面對的狀況等內容。

❹ 結尾（Closing）
於本文提出具體的公告內容後，最後再次提醒讀者必須銘記的內容，或是以致謝作結。

01 **it is our policy to...**　我們的方針是…
It is our policy to keep all personal information private.
我們的方針是將所有個人資訊保密。

02 **you are advised to...**　建議…（應知悉）
You are advised to report any discrepancies on the website.
建議您回報網站上任何不符之處。

03 **be on the lookout for...**　對…保持警惕、仔細觀察…
Please **be on the lookout for** any suspicious activity in the area.
請留意該地區任何可疑的活動。

04 **close for maintenance**　關閉維修
The Shanghai plant was **closed for regular maintenance**.
上海廠因爲定期維修暫時關閉。

05 **thank you for your support**　謝謝您的協助
Thank you for your support the last two years.
謝謝您過去兩年的協助。

06 **apologize for the inconvenience**　抱歉造成不便
We **apologize for the inconvenience** the downtime may have caused.
對於停機造成的不便，我們深感抱歉。

07 **until further notice**　在進一步公告之前
Please hold on to your receipt **until further notice**.
在收到進一步通知前，請保留您的發票。

08 **we invite you to...**　我們邀請您…
We invite you to visit our newest theme park.
我們歡迎您蒞臨最新的主題樂園。

小測驗

01. This store has been (   ).
　　這家店因為維修暫時關閉。

02. We sincerely (   ) the inconvenience.
　　造成不便，我們深感抱歉。

03. It (   ) to keep personal information private.
　　我們的方針是個人資訊保密。

04. The service will be suspended (   ).
　　在進一步公告之前將停止服務。

Temporary Closing of Tmart Superstore

This location will close for maintenance and upgrades beginning today, June 15, 20×× at 7 P.M. At this time, we do not have a reopen date. We thank you for your support and apologize for the inconvenience.

Until further notice, we invite you to shop with us at one of our convenient nearby locations.

Tmart Superstore #344
3349 Hillcrest Lane
Irvine, CA 92718

Tmart Superstore #348
1313 Prospect Street
Irvine, CA 92718

Tmart Superstore #351
3261 Goldie Lane
Irvine, CA 92718

單字

- temporary 暫時的
- close 關門、關閉
- location 地點、場所
- maintenance 維修、保養
- at this time 現在
- reopen 重新開放
- support 協助、支持
- apologize 道歉
- inconvenience 不便
- nearby 附近的
- further notice 日後公告
- invite 邀請、招待

提案計畫書

▶ 翻譯請參閱 p.216

 文書範本 1 商務提案計畫書

Business Proposal

Prepared for: T-Motors located at 1657 Eagle Drive, Southfield, MI
Prepared by: Spink Tire Company located at 3329 Bayo Drive, Joplin, MI

Summary

The purpose of this proposal is to forge a strategic business partnership between T-Motors and Spink Tire Company.

Introduction

Spink has developed quality automobile tires for over 100 years. We are known to manufacture tires with superior specifications and advantages compared to similar products in the industry.

Product and Service Overview

Spink specializes in products for coupé and sedans. Spink tires combine sporty looks, long duration and all-weather traction. The tires are especially well suited for wet and snowy roads.

Pricing

Size	Price (Per tire)
185/60R15	$119.45
185/65R15	$123.45
185/70R15	$127.45

Benefits

• All tires are certified by the SDA.
• Warranty of 6 years / 60,000 miles.
• 20% discount for bulk orders of 50 or more.

單字

• forge 建立、構築
• strategic 戰略性的
• business partnership 商業夥伴
• quality 品質優良的
• manufacture 製造
• superior 優秀的
• specialize in... 專門從事…
• traction 靜止摩擦力
• certify 證明
• pricing 定價、價格
• benefit 優惠
• bulk order 大量訂購、大宗訂單

美國百大企業文書撰寫祕訣！

Proposal（提案計畫書）的撰寫內容，是在向潛在客戶介紹可提供的服務或商品，希望對方採用。也就是說，目的在於向對方展現自己是特定領域的翹楚，吸引對方的青睞。提案大致上可分為三步驟。1) 先拋出爭議問題，或與客戶需求相關的內容。2) 接著詳細說明自己企劃的商品及服務。3) 最後報上提案商品及服務的價格。此外再加上資格、優惠、期間、其他費用、保險等具體條件，即可完成提案計畫書。

實務負責人的一句話！

撰寫提案計畫書之前，最好先確切掌握客戶的狀況後，再擬訂吻合需求的對策。如果能提出自己公司比競爭者更出色的原因來吸引對方的話，會更加分。

文書架構詳細解析

Business Proposal

❶ **Prepared for:** T-Motors located at 1657 Eagle Drive, Southfield, MI
Prepared by: Spink Tire Company located at 3329 Bayo Drive, Joplin, MI

Summary

❷ The purpose of this proposal is to forge a strategic business partnership between T-Motors and Spink Tire Company.

Introduction

❸ Spink has developed quality automobile tires for over 100 years. We are known to manufacture tires with superior specifications and advantages compared to similar products in the industry.

Product and Service Overview

❹ Spink specializes in products for coupé and sedans. Spink tires combine sporty looks, long duration and all-weather traction. The tires are especially well suited for wet and snowy roads.

Pricing

❺

Size	Price (Per tire)
185/60R15	$119.45
185/65R15	$123.45
185/70R15	$127.45

Benefits

❻ • All tires are certified by the SDA.
• Warranty of 6 years / 60,000 miles.
• 20% discount for bulk orders of 50 or more.

---

❶ 標題（Heading）
寫明提案計畫書的題目及收件人。也可寫出撰寫日期。

❷ 概要（Summary）
概述提案計畫書的目的及內容。內容以客戶要求（需求）及客戶目前關注的議題為中心。

❸ 介紹（Introduction）
此欄位介紹撰寫人的公司。簡述公司沿革及公司成就等資訊，積極宣傳公司的資訊。

❹ 產品和服務（Product and Services）
說明撰寫人想提供的產品或服務。

❺ 價格（Pricing）
列出產品和服務的價格。

❻ 優惠（Benefits）
提出產品與服務的相關優惠。

撰寫文書所需的主要用語

01 **the purpose of this proposal is...** 本提案計畫書的目的是…
The purpose of this proposal is to introduce a new line of toys.
本提案計畫書的目的是介紹新系列的玩具。

02 **forge a strategic partnership with...** 與…建立策略合作關係
You'll be interested to **forge a strategic partnership with** us.
相信貴公司會有興趣與本公司建立策略合作關係。

03 **provide customers with...** 向客戶提供…
The new device **provides customers with** fast Wi-Fi connection.
新設備提供客戶高速 Wi-Fi 連線服務。

04 **compared to similar products in the industry** 與業界類似產品比較
The T2 is inexpensive **compared to similar products in the industry**.
T2 的價格比業界類似產品低廉。

05 **specialize in...** 專精於…
We **specialize in** over-the-counter drugs.
成藥是我們的專門領域。

06 **be well suited for...** 會很適合…
This project **is well suited for** Malaysia's tropical environment.
本計劃十分適合馬來西亞的熱帶環境。

07 **offer a variety of...** 提供各式各樣的…
The organization **offers a variety of** health and human services.
本單位提供各式各樣的健康福利服務。

08 **be covered in the package** 包含於商品中
Delivery **is** not **covered in the package**.
商品不含配送服務。

小測驗

01. We ( ) of memory chips.
    本公司提供各式各樣的記憶體晶片。

02. We are interested in ( ) with you.
    本公司有興趣與貴公司建立策略合作關係。

03. Tiema Co. ( ) car rentals.
    梯馬達公司專營汽車租賃。

04. ( ) in the industry the GX1 is very sturdily built.
    與業界類似產品相比，GX1 非常堅固。

European Business Travel Proposal

Prepared for: Brandon Parker, Keithmore Trading Company
Prepared by: Randal Lee, ADS Travel Agency

About Us

ADS is a licensed travel agency located in Houston, Texas. We offer a variety of travel packages for companies, individuals and groups.

Package Overview

Our business travel package covers all countries in Europe including Turkey and the Mediterranean Sea islands. Trips may be as long as five days and use all types of transportation including airplane, train and ship.

Pricing (Round trip)

Travel class	Price (Individual)	Price (Groups of 4 or more)
Economy	$2,150.00	$1,705.00
Business Class	$4,225.00	$3,805.00
First Class	$7,400.00	$6,555.00

Benefits

Travelers can choose their itinerary on our website. Options such as hotels and means of transportation can also be selected. Additional luxuries such as entertainment are not covered in the package.

Insurance

Travel insurance is paid for by the agency and covers cancellations, lost luggage, medical emergencies, and other unforeseen emergencies that might arise during your trip.

單字

- licensed 有資格的、被許可的
- overview 概要
- travel package 旅遊方案
- cover 處理、包含、承保
- including... 包含⋯
- Mediterranean Sea 地中海
- round trip 往返旅行
- itinerary 旅行日程表
- option 選擇、選項
- luxury 奢侈
- travel insurance 旅行保險
- unforeseen 意料之外的

說明書

▶翻譯請參閱 p.218

 文書範本 1 流程說明書

Procedure Manual – Annual Report

1.0 Purpose and Scope

1.1 This manual describes the process of preparing and submitting the annual report to the CEO at Meyer Corporation.

2.0 Responsibility

2.1 Chief responsibility of the task falls on the following:

2.1.1 Chief Operating Officer (COO)

2.1.2 Chief Financial Officer (CFO)

2.1.3 Marketing Director

3.0 Definitions

3.1 Annual Report – Comprehensive financial report is issued to the CEO every January that lists the company's annual sales and profit.

4.0 Reference

4.1 20×× Annual Report

4.2 Annual Report Format

4.3 Invoice Register

5.0 Procedure

5.1. The annual report is prepared by the COO, in accordance with the requirements defined by the CEO.

5.2. The annual report uses the already established annual report format.

5.3. The COO and Marketing Director, after reviewing the report, must both approve it by the deadline of January 31 of each year.

Note: The annual report covers the period from January 1 to December 31 of the previous calendar year.

單字

- annual report 年度報告
- scope 範圍
- describe 描述、說明
- chief responsibility 主要職責
- task 工作、任務
- fall on... 承擔（責任、義務等）

- Chief Executive Officer (CEO) 執行長
- Chief Operating Officer (COO) 營運長
- Chief Finance Officer (CFO) 財務長
- invoice register 發貨單記錄本
- approve 批准
- note 備註

美國百大企業文書撰寫祕訣！

Manual（說明書）是指為特定流程及工作執行、服務及產品的使用提供指南的文書。目的可能是政策（policy）、程序（procedure）、事業營運（operation）、教育訓練（training）等，其中以 Policy and Procedure（政策及程序）及 Instruction Manual（使用說明書）最常見。所有說明書的目標都是傳達正確的指引，因此寫作時的重點是要方便讀者理解。

實務負責人的一句話！

撰寫時，最好以讀者的角度撰寫，以主動句呈現說明書內容。說明書的目的是協助讀者達成特定目標或執行任務，所以寫作應簡明扼要，避免提供不必要的資訊。

文書架構詳細解析

---

❶ Procedure Manual – Annual Report

1.0 Purpose and Scope
❷ 1.1 This manual describes the process of preparing and submitting the annual report to the CEO at Meyer Corporation.

2.0 Responsibility
2.1 Chief responsibility of the task falls on the following:
2.1.1 Chief Operating Officer (COO)
2.1.2 Chief Financial Officer (CFO)
2.1.3 Marketing Director

3.0 Definitions
3.1 Annual Report – Comprehensive financial report is issued to the CEO every January that lists the company's annual sales and profit.

❸ 4.0 Reference
4.1 20×× Annual Report
4.2 Annual Report Format
4.3 Invoice Register

5.0 Procedure
5.1. The annual report is prepared by the COO, in accordance with the requirements defined by the CEO.
5.2. The annual report uses the already established annual report format.
5.3. The COO and Marketing Director, after reviewing the report, must both approve it by the deadline of January 31 of each year.

❹ Note: The annual report covers the period from January 1 to December 31 of the previous calendar year.

---

❶ 標題（Heading）
記載說明書的目的、撰寫日期，以及提供說明書的公司名稱（單位名稱）。

❷ 目的（Purpose）
明確概述說明書整體的目的或主旨。最好以主要關鍵字為中心撰寫內容，以利傳達說明書的核心主旨。

❸ 本文（Main Body）
記載說明書的核心流程及方針內容。同時本文中，也可以納入說明方針的適用對象及說明書的定義，或是其他值得參考的資料等。

❹ 備註（Note）
註記本文中未能提及的重要事項等。舉例來說，產品上標示的資訊和說明書的適用日期等都可以記載於備註欄。

01 **describe the process of...** 說明⋯的流程（過程）
The following will precisely **describe the process of** running SDK 2.0.
以下將精確說明 SDK 2.0 的執行流程。

02 **be prepared by...** 由⋯準備
The agenda is to **be prepared by** the minutes taker.
議程將由會議記錄撰寫人準備。

03 **responsibility of... falls on...** 由⋯負責
The **responsibility of** testing **falls on** the chief engineer.
由總工程師負責測試。

04 **in accordance with...** 依據⋯
In accordance with company rules, the store must close at 7:30 P.M.
依公司規定，商店應於晚上 7:30 打烊。

05 **cover the period from... to...** 從⋯到⋯有效
The warranty **covers the period from** May 1, 20✕✕ **to** April 30, 20✕✕.
保固期為 20✕✕ 年 5 月 1 日至 20✕✕ 年 4 月 30 日。

06 **provide guidelines to...** 向⋯提供指南
This user guide **provides guidelines to** users of the security program ePrinter.
此使用手冊為安全程式 ePrinter 的用戶提供指南。

07 **is obligated to...** 有⋯的義務
The manager in charge **is obligated to** hand out instructions.
負責的管理人有義務做出指示。

08 **be limited to...** 只限於⋯
Promotional activities will **be limited to** the immediate vicinity.
促銷活動只限於鄰近地區。

小測驗

01. This manual was ( ) Ken Lehman.
    這本手冊是由肯・雷曼準備的。

02. You will ( ) to check the machines daily.
    你有義務每天確認機器。

03. The following ( ) of contract writing.
    以下將說明契約書撰寫的流程。

04. ( ) with company policy, work hours are from 9 A.M. to 6 P.M.
    依公司規定，工作時間從早上 9 點到晚上 6 點。

Policy and Procedure Manual
St. Joseph's Medical Center
Kansas City, Missouri

Policy No.: 133
Subject: Handling and dosing of K-101
Purpose: To provide guidelines to pharmacists regarding the dosing of the drug K-101 to eligible patients.

Policy and Procedure:

I. Eligible patients
   - Is 18 years old or older.
   - Is a patient at St. Joseph's Medical Center.
   - Has written a prescription specific to K-101.

II. Procedure when a K-101 prescription is received:
   - Pharmacist will verify the prescription by checking the central database.
   - Pharmacist will hand out the exact amount of the drug as described in the prescription.
   - Pharmacist will convey clear instructions to the patient regarding the taking of the drug.
   - Pharmacist will be obligated to provide additional doses needed by the patient after the original amount is consumed.

III. Prescriptions Limits
   - Patients are limited to 6 tablets a day, 42 tablets a week unless there's a special need recommended by the physician.

Author: Dr. Eric M. Blakely
Effective Date: March 11, 20××
Revised Date(s): April 27, 20××, October 9, 20××

單字

- pharmacist 藥劑師
- dose 給…用藥、劑量
- guideline 準則
- eligible 有資格的、有條件的
- prescription 處方箋
- verify 證明、證實
- additional 附加的
- consume 消耗
- tablet 錠、顆
- recommend 建議、勸告
- effective date 生效日期
- revised date 修正日期

文書範本 1 公司內部電子報　　　　　　　　▶翻譯請參閱 p.220

Kesco Newsletter
- March 20×× Issue -

New Launch!

INOS 2.0 which was released on March 1, 20×× is a huge upgrade over INOS 1.0 with significant bug fixes and brand new functions. INOS which stands for Inter-Multitasking Operating System is a multitasking system which allows users to manage various tasks in an all-in-one solution. Click here to find out more.

Employee of the Month

March's employee of the month is Rachel Shin from the software research center. Her dedication and constant efforts in the development of INOS 2.0 for the last two years has been invaluable in its successful launch. The launch is expected to garner much attention and investment to the multitasking field.

Upcoming Events

Wednesday, March 8: Launching of cloud service system Kesco PRL
Friday, March 17: Annual Kesco Developer Conference (Kesco Headquarters)
Monday, March 20: Release of Patch 4.5 for Ureka System

Virus Alert!

A malicious virus has been making the rounds in the last week of February. Be on the alert for emails titled "Outstanding invoice" which is known to have a suspicious attachment and "Amazing New Stamina Drug!" which has been verified as a harmful phishing email.

Kesco Public Relations Team
Frederick Creech | 070-3736-1100 | pr@kesco.com

單字

- launch 發行、上市
- release 上市
- significant 重要的、重大的
- bug fix 修復錯誤
- brand new 全新的
- stand for... 代表…、…的縮寫
- all-in-one solution 統一的解決方法
- dedication 奉獻
- constant 連續不斷的
- invaluable 非常有用的、無價的
- garner 獲得、收集
- upcoming 即將發生的
- suspicious 可疑的
- harmful 有害的

美國百大企業文書撰寫祕訣！

Newsletter（電子報）是企業、機構及各種團體定期向團體成員發送的刊物。電子報的內容包括近期活動、新消息、會議相關資訊等。電子報一般是發行給機構內的成員，不過企業以宣傳為目的發送給潛在客戶的信件也是電子報的一種。過去電子報以印刷品的形式發送，不過近年來利用電子郵件及電子檔的形式發布較為普遍。

實務負責人的一句話！

製作吸引人的內容及引發讀者的興趣是製作電子報的重點，因此電子報經常會搭配繽紛的圖片與搶眼的排版。

文書架構詳細解析

❶
Kesco Newsletter
- March 20×× Issue -

New Launch!
INOS 2.0 which was released on March 1, 20×× is a huge upgrade over INOS 1.0 with significant bug fixes and brand new functions. INOS which stands for Inter-Multitasking Operating System is a multitasking system which allows users to manage various tasks in an all-in-one solution. Click here to find out more.

Employee of the Month
March's employee of the month is Rachel Shin from the software research center. Her dedication and constant efforts in the development of INOS 2.0 for the last two years has been invaluable in its successful launch. The launch is expected to garner much attention and investment to the multitasking field.

❷

Upcoming Events
Wednesday, March 8: Launching of cloud service system Kesco PRL
Friday, March 17: Annual Kesco Developer Conference (Kesco Headquarters)
Monday, March 20: Release of Patch 4.5 for Ureka System

Virus Alert!
A malicious virus has been making the rounds in the last week of February. Be on the alert for emails titled "Outstanding invoice" which is known to have a suspicious attachment and "Amazing New Stamina Drug!" which has been verified as a harmful phishing email.

❸
Kesco Public Relations Team
Frederick Creech ｜ 070-3736-1100 ｜ pr@kesco.com

❶ **標題（Heading）**
記載電子報的標題、副標題、刊號、日期等。

❷ **本文（Main Body）**
依類型需求，電子報可以加入形式豐富的多樣化內容。舉例來說，如為公司內部通訊，內容可以收錄公司開發的新產品、當月優秀職員等近期新聞。對客戶發布的電子報則會提供新產品及各種活動的資訊。另外，本文內容大多會插入各種圖片，以激起讀者的興趣。

❸ **結尾（Closing）**
電子報不需要特別的結尾，建議提供發行人（單位）的聯絡方式，以便讀者提出回饋或疑問。

撰寫文書所需的主要用語

01 **be a huge upgrade over...** …大大升級
This new version of the program **is a huge upgrade over** the previous one.
程式的新版本對前一版本做出了重大升級。

02 **allow users to...** 讓使用者可以…
This new function **allows users to** freely communicate with each other.
這項新功能讓使用者能夠自由地相互交流。

03 **be expected to garner...** 預計將獲得…
The festival **is expected to garner** plenty of attention.
這場慶典預計將獲得許多關注。

04 **make the rounds** 巡迴、流傳
The sign-up sheet is currently **making the rounds**.
報名表目前正在人們之間傳遞。

05 **be known to have...** 以…而為人所知
The new facility **is known to have** a gym open to all employees.
這座新設施以有提供給員工使用的健身房而廣為人知。

06 **be verified** 經驗證（經公開）
The promotion has **been verified** to be a hoax.
那個促銷活動已被證實是一場騙局。

07 **this edition contains...** 這一期包含…
This edition contains numerous tips on how to use the Atrex machine.
本期收錄了許多使用阿特雷克斯機器的小祕訣。

08 **offer a free trial** 提供免費體驗
Amazon is now **offering a free trial** of its Air Premium service.
亞馬遜現在正在提供 Air Premium 服務的免費體驗。

小測驗

01. The new theme park is (　　　　　　　) huge investments.
新的主題樂園預計將獲得巨額投資。

02. This (　　　　　　　) numerous articles on the event.
這一期收錄了許多有關活動的報導。

03. Version 3.0 is a huge (　　　　　　　) the previous one.
3.0 版本對前一版本做出了重大升級。

04. We are (　　　　　　　) trial of our services.
本公司正在提供免費體驗我們的服務。

The Professional

- Monthly newsletter for users of Pro Office 480 -

March 11, 20××

A Word From the Editor

Hello and welcome to the fifth edition of The Professional. This edition contains the latest information and advice on the cloud feature of Pro Office.

In This Issue

- Schedule Management on Pro Office 480
- Feature of the Month: Text Clouding
- Upcoming Updates
- Testimonial Section: IT Analyst Max Brown

Schedule Management? Easy Right? Make It Even Easier!

Schedule management has never been easier thanks to Work Manager in Pro Office 480. Work Manager is the brainchild of Marty Jones who worked on the program for more than three years. It offers convenient interconnection with other Pro Office 480 programs. Read More.

Free Trial!

The Professional is offering a free one month trial of **Community Help** to all registered members. Through **Community Help** members may use a 24 hour online help service if you have any questions or problems using Pro Office 480. Read More.

Unsubscribe

The Professional
Email: editor@theprofessional.com | Website: www.theprofessional.com

單字

- edition 版本
- contain 包含、含有
- advice 建議、忠告
- feature 特徵、特色
- testimonial 推薦文、推薦
- thanks to... 由於…、多虧…

- brainchild 點子、發明
- convenient 方便的
- interconnection 相互連接、連動
- free trial 免費體驗
- registered member 已註冊的會員
- unsubscribe 取消訂閱

文書範本 1　公司文宣　　　　　　　　　▶翻譯請參閱 p.222

Is Your Banking Safe?

Cyan Bank
Online Banking Done Right

Cyan Bank History

19×× Founded as Cyan Banking Co.
19×× Opens country's first online banking system
19×× Enters Chinese market
20×× Enters Indian market
20×× Celebrates 40th anniversary

Why Us?

Cyan Bank is one of America's biggest, most trusted banks. We specialize in safe, hack-proof online banking. Through our online and mobile banking system you can process safe bill payments, account transfers and fund transfers.

Questions?

Visit a Cyan Bank branch near your or visit www.cyanbank.com for more information.

Cyan Bank Corporate Center
2807 Columbia Boulevard
Windsor Mill, MD 21244

www.cyanbank.com

單字

- found 設立、建立
- enter 進入
- market 市場
- celebrate 紀念
- anniversary 紀念日
- trusted 值得信賴的、可信的
- specialize in... 專門從事…
- hack-proof 防駭客的
- process 處理
- bill payment 帳單繳費
- account transfer 轉帳
- fund transfer 資金轉移

美國百大企業文書撰寫祕訣！

Brochure（文宣）是公司或機構基於行銷目的，將自家產品及服務的相關內容整理成傳單發放的文書。這種文宣可能是一本小手冊，也可能是一張紙而已。目前依然有很多以印刷品形式製作的文宣，不過像電子信一樣，在網頁上或以電子郵件形式發布的文宣也很常見。文宣的內容組成及可以傳達的內容多變，製作時務必要做好企劃和設計。

實務負責人的一句話！

讀者看到文宣之後，一般來說會在 5 秒內決定要讀或不讀內容。因此要設計令人印象深刻的標題、圖片等，例如加上公司的代表圖像或標語來吸引讀者的注意。

文書架構詳細解析

---

❶

Is Your Banking Safe?
Cyan Bank
Online Banking Done Right

Cyan Bank History

19×× Founded as Cyan Banking Co.
19×× Opens country's first online banking system
19×× Enters Chinese market
20×× Enters Indian market
❷ 20×× Celebrates 40th anniversary

Why Us?

Cyan Bank is one of America's biggest, most trusted banks. We specialize in safe, hack-proof online banking. Through our online and mobile banking system you can process safe bill payments, account transfers and fund transfers.

Questions?

Visit a Cyan Bank branch near your or visit www.cyanbank.com for more information.

❸

Cyan Bank Corporate Center
2807 Columbia Boulevard
Windsor Mill, MD 21244

www.cyanbank.com

---

❶ 標題（Heading）
撰寫時應包含公司名稱、公司的代表圖像及標語、產品圖片，用吸引人的內容一下子抓住讀者的注意力。

❷ 本文（Main Body）
本文中應列出公司歷史、公司提供的服務，還有公司的強項等。這部分要活用各種圖片或照片、圖標等，盡可能抓住讀者的視線。此外，本文的內容鋪陳也要流暢，才能充分誘使讀者產生興趣。

❸ 結尾（Closing）
記載讀者可聯絡的電話號碼、官網網址、電子信箱等資訊。也可一併提供產品購買或使用服務的管道。

撰寫文書所需的主要用語

01 **founded as...** 創立⋯
Founded as SUN Studio in June 20××.
20×× 年 6 月太陽工作室成立。

02 **open the country's first...** 開辦全國第一個⋯
The Indian government **opened the country's first** e-court.
印度政府開辦全國第一個線上法庭。

03 **visit... for more information** 前往⋯以獲得更多資訊
Visit us at www.americancard.com **for more information**.
如欲獲得更多資訊，請至 www.americancard.com。

04 **begin expansion in...** 在⋯擴張、進軍⋯
The company **began expansion in** the Western United States in 19××.
該公司於 19×× 開始進軍美國西部市場。

05 **celebrate... th anniversary** 慶祝（迎接）⋯週年
Alphabet Inc. just **celebrated** its **first anniversary**.
Alphabet 剛迎來公司成立一週年。

06 **be one of the most...** 最⋯的之一
WellPoint **is one of the most** dynamic companies in the world.
WellPoint 公司是全球最活躍的公司之一。

07 **It takes just...** 只需要⋯
It takes just three simple steps to send your shipment.
只需要 3 個簡單的步驟即可寄送您的貨物。

08 **check out our...** 來看看我們的⋯
Check out our newest book now.
立刻來看看我們的新書吧。

小測驗

01. Visit us at www.realster.com (   ).
如需獲得更多資訊，請至 www.realster.com。

02. RaMX is (   ) original IT companies in Korea.
RaMX 公司是韓國最具原創性的科技公司之一。

03. (   ) our newest product now.
立刻來看看新產品吧。

04. Trigest (   ) its 10th anniversary in 20××.
崔葛斯特公司在 20xx 年慶祝成立 10 週年。

Why Ship with JTS?

JTS Online Express

Simple, Quick and Dependable

It's Simple

It takes just three simple steps to send your shipment. All online!

It's Dependable

Delivery without loss or damage is guaranteed. Otherwise we compensate you 100%!

It's Quick

Expect your package to be picked up in less than 24 hours and to arrive in less than 24 hours. 48 hours in total!

It's Convenient

Tracking your packages has never been easier! Check out our real-time tracking system at the JTS Express website!

How to get started?

1. Go to www.jtsexpress.com
2. Register.
3. Make an order!

單字

- ship 運送、輸送
- shipment 運送的貨物
- pick up 領取
- dependable 可信的、可靠的
- loss 損失
- damage 損壞

- otherwise 否則
- compensate 賠償、補償
- tracking 追蹤
- check out 查看
- real-time 即時
- register 註冊、加入會員

電話留言

 文書範本 1 電話留言 ▶ 翻譯請參閱 p.224

Sherman Energy Co.

Telephone Message

For: *Kelly Kim*

Caller Information

☑ **Mr.** ☐ **Ms.** *Joseph High*
Company: *TraderFarm Co.*
Phone No.: *(562) 344-3562*

☑ Please Call   ☐ Will Call Again
☐ Returned Your Call   ☐ Called to See You
☑ Left a Message (below)

Message:
Called to check on the September 23 order. Wants a return call by today.

Date: *Sep. 25, 20✕✕*   **Time**: *3:05 P.M.*
Signed: *Pat Ames*

單字

- telephone message 電話訊息
- caller 打電話的人、發信者
- leave a message 留下的訊息
- return a call 回電

美國百大企業文書撰寫祕訣！

Phone message（電話留言）是電話無人接聽時留給致電對象的訊息。如果沒有特定的表格，也可以用紙張、通訊軟體或簡訊留下訊息。如果已有特定的表格，就只須填入致電者的姓名、公司名稱、及欲傳達的訊息即可。若想要更正式有禮地留下完整訊息，也可以詳細寫出致電者的聯絡方式、致電時間與致電對象的姓名。

實務負責人的一句話！

撰寫電話留言時，為了方便致電對象了解狀況，應正確寫出致電者姓名、公司名稱、致電目的、致電者的聯絡方式等資訊，並且寫明希望致電對象採取的回應方式（例如回電等）。

文書架構詳細解析

Sherman Energy Co.

Telephone Message

❶

❷ **For:** *Kelly Kim*

Caller Information

❸ ☑ **Mr.** ☐ **Ms.** *Joseph High*
Company: *TraderFarm Co.*
Phone No.: *(562) 344-3562*

☑ Please Call ☐ Will Call Again
☐ Returned Your Call ☐ Called to See You
☑ Left a Message (below)

❹ **Message:**
Called to check on the September 23 order. Wants a return call by today.

❺ **Date:** *Sep. 25, 20XX* **Time:** *3:05 P.M.*
Signed: *Pat Ames*

❶ **標題（Title）**
文書標題寫為 Telephone Message 或 Message。如已有既定格式，標題也可以直接寫公司名稱就好。

❷ **致電對象（Recipient）**
寫出致電對象的姓名。

❸ **致電者資訊（Caller Information）**
留下致電者的姓名、公司、電話號碼等。

❹ **留言內容（Message）**
寫出欲傳達給致電對象的內容。記載內容時要明確寫出致電目的及希望致電對象採取的應對措施。

❺ **結尾（Closing）**
留下致電時間及電話留言經手人的姓名。

01 **return a call**　回電
Brandon Gill **returned your call**.
布蘭登·吉爾回電了。

02 **leave a message**　留言
Mr. Simmons **left a message** for you.
賽蒙思先生有留言給你。

03 **call to check on...**　打電話確認…
Vice President Wallace **called to check on** the project.
華萊士副會長打電話來確認計畫狀況。

04 **call for someone...**　找…
Mr. Murray **called for you**.
默里先生在找你。

05 **call back**　回電
She **called back** to confirm the meeting.
她回電確認會議的事情。

06 **message taken by...**　訊息記錄者為…
Message taken by Dong-nam Shin.
訊息記錄者為申東南。

07 **meet to discuss...**　見面討論…
He would like to **meet to discuss** prices.
他想要見面討論價格。

08 **make an appointment**　預約（開會）時間
Christy Yamaguchi would like to **make an appointment** with you.
克里斯蒂·山口想跟你預約開會時間。

小測驗

01. Mr. Dennis (　　　　　　　　　　) you.
丹尼斯先生在找你。

02. Douglas (　　　　　　　　　　) for you this morning.
道格拉斯今天早上有留言給你。

03. He (　　　　　　　　　　) to confirm something.
他回電確認事情。

04. Jane Kang would like to (　　　　　　　　　　).
簡·康想約個時間。

MESSAGE

FOR *Alonzo Gomez*
DATE *July 5, 20✕✕*   TIME *11:20* (A.M.)/ P.M.
FROM *Juan Aguila*
PHONE *(311) 494-5561*

- [✓] Called For You
- [ ] Returned Your Call
- [✓] Needs to See You
- [ ] Sent an Email

- [ ] Please Call Back
- [✓] Will Call Back
- [ ] Urgent
- [ ] Sent a Fax

MESSAGE

Would like to meet as soon as possible to discuss contract terms. Will call again this afternoon to make an appointment.

TAKEN BY *Misty Kapoor*

單字

- call back 回電
- urgent 緊急的、急迫的
- as soon as possible 盡快

- discuss 討論
- contract terms 合約條款
- appointment 約定

新聞稿

▶ 翻譯請參閱 p.226

文書範本 1 新聞稿（新產品）

Key Widget Unveils Revolutionary Tablet Series: Interactive Pro

LOS ANGELES, August 24, 20×× – Key Widget introduced today a new tablet computer called "Interactive Pro," the first generation of its kind. It features a groundbreaking multi-touch display and a seventh-generation quad-core processor.

The device was introduced at the LA Technology Expo by CEO Janet Hillman. "It's fitting that this all-new generation tablet series is our proudest achievement yet," she said. "We are confident that consumers will love its ease of use and unique design."

The Interactive Pro offers the following features:

• Multi-touch screen three times more sensitive than the previous Interactive X
• Seventh-generation quad-core processor
• 550 g in weight
• Option of 128 or 256 GB storage capacity

For additional specifications, custom options and accessories visit www.keywidget. com/interactivepro.

About Key Widget

Headquartered in Torrance, California, Key Widget specializes in bringing groundbreaking high-tech products to your home.

Press Contact

Laura Hayes
lauraehayes@keywidget.com
(901) 763-9146

單字

• revolutionary 劃時代的
• groundbreaking 開創性的
• device 裝置、機器
• fitting 適合的
• achievement 成就
• unique 唯一的、獨特的

• sensitive 敏感的
• storage capacity 儲存容量
• specification 規格
• headquarter 總部設於…
• custom option 客製化選項
• accessory 配件

美國百大企業文書撰寫祕訣！

Press release（新聞稿）是機構或企業向媒體發表的文書。企業為了宣傳，將新聞稿發給報紙、雜誌、廣播、新聞節目、網路媒體等媒體單位，媒體記者會參考新聞稿撰寫報導，或是潤飾過後再發布。公司發新聞稿的目的是自我宣傳，所以撰寫新聞稿時要想辦法吸引讀者的關注。除了透過媒體，企業也會直接在自己的官方網站上發表新聞稿。

實務負責人的一句話！

如果想提高新聞稿內容出現在媒體上的可能性，就要盡可能以符合新聞稿的格式撰寫。可以參考下列格式，從標題到結尾仔細編寫。

文書架構詳細解析

❶ Key Widget Unveils Revolutionary Tablet Series: Interactive Pro

❷ LOS ANGELES, August 24, 20×× – **❸** Key Widget introduced today a new tablet computer called "Interactive Pro," the first generation of its kind. It features a groundbreaking multi-touch display and a seventh-generation quad-core processor.

The device was introduced at the LA Technology Expo by CEO Janet Hillman. "It's fitting that this all-new generation tablet series is our proudest achievement yet," she said. "We are confident that consumers will love its ease of use and unique design."

The Interactive Pro offers the following features:

❹ • Multi-touch screen three times more sensitive than the previous Interactive X
• Seventh-generation quad-core processor
• 550 g in weight
• Option of 128 or 256 GB storage capacity

For additional specifications, custom options and accessories visit www.keywidget.com/interactivepro.

About Key Widget

Headquartered in Torrance, California, Key Widget specializes in bringing groundbreaking high-tech products to your home.

❺ Press Contact

Laura Hayes
lauraehayes@keywidget.com
(901) 763-9146

❶ 標題（Title）
選對合適的標題能夠有效反映出想傳達的新聞內容，一定要想出最能夠吸引讀者目光的標題。

❷ 日期（Dateline）
記載所在城市名稱及發表日期。

❸ 前言（Introduction）
用能夠回答 5W（who, what, where, when and why）的內容寫出第一段。以盡可能符合 5W 要素為書寫原則，將能更精準地傳達出核心內容。

❹ 本文（Main Body）
活用事發背景、證據資料、引用發言等要素，盡量詳細呈現欲傳達的新聞內容。

❺ 聯絡方式（Contact Information）
留下機構資訊及撰寫者聯絡方式（電話號碼和電子信箱等）。

撰寫文書所需的主要用語

01 **be introduced at...** 在…上亮相
Ford's newest SUV **was introduced at** the Detroit Car Show.
福特公司的最新款休旅車在底特律車展上亮相。

02 **be the first of its kind** 首次出現在該領域
The 4D TV Smartbox **is** truly **the first of its kind**.
4D 電視機上盒首次出現在該領域。

03 **offer the following features** 提供下列功能（特點）
The system **offers the following features**.
這個系統提供下列功能。

04 **for additional specifications** 為了查看更多規格
Click this link **for additional specifications**.
如欲查看更多規格，請點擊此連結。

05 **have a long history of...** 有…的悠久歷史
Shell **has a long history of** safe drilling operations.
殼牌公司有安全鑽井作業的悠久歷史。

06 **be one's specialty** 是…的專業
Electronic manufacturing **is Foxconn's specialty**.
製造電子產品是富士康公司的專業領域。

07 **employ... associates** 雇有…的員工
Prudential Financial **employs** more than 48,000 **associates** worldwide.
保誠金融公司在全球雇有超過 48,000 名員工。

08 **to learn more about...** 進一步瞭解…
To learn more about this special deal, visit our website.
如果想知道更多關於特價商品的資訊，請見本公司網站。

小測驗

01. The new phone (   ) at the Mobile World Congress.
新手機在世界行動通訊大會上亮相。

02. Kroger has (   ) of excellent customer service.
克羅格公司有卓越客戶服務的悠久歷史。

03. Pfizer (   ) more than 90,000 associates.
輝瑞公司雇有 9 萬名以上的員工。

04. The notebook (   ) features.
這個筆記型電腦提供下列功能。

Calhoun Logistics Commits $100,000 Towards Hurricane Relief in Florida

MIAMI, Florida – September 20, 20×× – Calhoun Logistics has committed to support the relief cause in the hurricane hit area of south Florida with $100,000 in cash and relief aid.

Hurricane Linda hit the area hard on the night of September 18 destroying thousands of homes and injuring hundreds. "We are committed to doing anything we can to bring relief to those affected," said Eric Flores, Senior Vice President of Operations.

Calhoun Logistics has a long history of providing critical aid in times of disaster. Transporting supplies has been Calhoun's specialty which was especially helpful in the 20×× Hurricane Katrina disaster when it delivered 10,000 packages of food, medicine and clothes to victims.

About Calhoun Logistics

Calhoun Logistics is a global logistics company that delivers millions of packages every day throughout the world. Calhoun has hub centers in 25 countries and employs 900,000 associates around the world. To learn more about Calhoun Logistics, visit www.calhoun.com.

Contacts

Ronald Kaczmarek
(542) 484-7130
pr1@calhoun.com

Carla Underwood
(542) 484-7131
pr2@calhoun.com

單字

- commit 承諾、投入（時間或金錢）
- relief 救濟、救濟品
- aid 援助、支援
- hit 席捲、打擊
- destroy 破壞
- critical 重大的
- disaster 災難
- supplies 物資、物品
- logistics 物流
- throughout the world 全世界
- hub center 物流倉儲
- associate （事業、職場上的）同事、員工

保固證明

 文書範本 1 保固證明 ▶ 翻譯請參閱 p.228

SD Product Warranty

This warranty applies to the SD Inkjet 5460 Printer.

Warranty Period

This warranty guarantees the provision of free service for a period of **12 months** from the purchase date stated on the warranty.

Warranty Terms and Conditions

The warranty covers any manufacturing defects and defects arising from normal usage within the 12 month warranty period.

If during the warranty period you submit a claim to SD in accordance with this warranty, SD will
1. repair the product using parts that are equivalent in quality and performance, OR
2. replace the product with the same model or, if you wish, a similar product with equivalent functionality, OR
3. exchange the product for a refund of your purchase price.

Obtaining Warranty Service

To obtain warranty service you are required to present the warranty card together with the purchase receipt for your free warranty service at a service center near you.

Limitation of Liability

This warranty does not apply to damages resulting from neglect, misuse, abuse, natural disaster, abnormal voltage or unforeseen circumstances such as wetting, contamination, lightning, etc. The warranty also does not cover damage due to alteration, adjustment or repair by an unauthorized person.

單字

- warranty 品質保證（書）
- warranty period 保固期間
- free service 免費服務
- defect 瑕疵
- normal usage 正常使用
- repair 修理
- equivalent 一樣的、同等的
- refund 退費
- receipt 收據
- neglect 不注意、疏忽
- natural disaster 自然災害
- unforeseen circumstance 意外情況
- alteration 變更
- unauthorized person 未經許可的人

美國百大企業文書撰寫祕訣 ！

Warranty（保固證明）是用來證明產品或設施品質的文書。若產品或設施在保固期間未能達到明定的標準，該機構保證會提供售後服務及系統性的維護修繕。保固證明內容要詳細填寫，以確保服務執行無礙。保證卡上應寫清楚產品名稱、型號、保固期等資訊。

實務負責人的一句話 ！

因消費者會在產品或設施於服務期間發生問題時，拿出保固證明來看，因此須鉅細靡遺的撰寫內容，以便提供切合消費者潛在需求的服務。

文書架構詳細解析

SD Product Warranty

❶ This warranty applies to the SD Inkjet 5460 Printer.

Warranty Period
❷ This warranty guarantees the provision of free service for a period of **12 months** from the purchase date stated on the warranty.

Warranty Terms and Conditions
The warranty covers any manufacturing defects and defects arising from normal usage within the 12 month warranty period.

❸ If during the warranty period you submit a claim to SD in accordance with this warranty, SD will
1. repair the product using parts that are equivalent in quality and performance, OR
2. replace the product with the same model or, if you wish, a similar product with equivalent functionality, OR
3. exchange the product for a refund of your purchase price.

Obtaining Warranty Service
❹ To obtain warranty service you are required to present the warranty card together with the purchase receipt for your free warranty service at a service center near you.

Limitation of Liability
❺ This warranty does not apply to damages resulting from neglect, misuse, abuse, natural disaster, abnormal voltage or unforeseen circumstances such as wetting, contamination, lightning, etc. The warranty also does not cover damage due to alteration, adjustment or repair by an unauthorized person.

❶ 標題（Heading）
將受保固的商品名稱加上 Warranty 寫出標題，再正確寫出保固證明所保障的產品型號／設施／服務的名稱。

❷ 保固期（Warranty Period）
明確記載保固期間，以粗體標示數字的部分，讓人一目了然。

❸ 條件（Terms and Conditions）
列出保固證明所規範的服務、適用條件、提供的服務種類。

❹ 保固服務（Warranty Service）
詳細敘述如何利用保固證明取得服務。

❺ 其他（Other）
寫出責任限制、使用者責任、附加資訊、其他事項等內容。

撰寫文書所需的主要用語

01 **this warranty applies to...** 這份保固證明適用於…
This warranty applies to all Mintchem products.
這份保固證明適用於明特秤公司所有的產品。

02 **guarantee the provision of...** 保證提供…
SE Enterprise **guarantees the provision of** free onsite repair.
SE 公司保證提供免費到府維修服務。

03 **repair a product** 維修產品
The manufacturer will **repair the product** within seven days.
製造商將在 7 天內維修產品。

04 **obtain warranty service** 獲得保固服務
To **obtain warranty service**, please visit the nearest SD service center.
如欲使用保固服務，請蒞臨離您最近的 SD 服務中心。

05 **you are required to...** 你必須…
You are required to pay shipping fees.
您必須支付運費。

06 **cover damage due to...** 保障因…受到的損傷
This warranty does not **cover damage due to** misuse.
本保固服務不保障因錯誤使用造成的損傷。

07 **replace a defective product** 更換瑕疵品
Star Appliances will **replace all defective products** made at its Chinese factory.
明星電器公司將更換其中國工廠製造的所有瑕疵品。

08 **ship a replacement product** 寄出替換用產品
We will **ship a replacement product** as soon as we receive your defective product.
一收到您提供的瑕疵品後，我們將會盡快寄出替換用產品。

小測驗

01. We (   ) of free repair.
    我們保證提供免費維修服務。

02. The warranty does not (   ) due to misuse.
    本保固服務不保障因使用錯誤造成的損傷。

03. KOP will (   ) product.
    KOP 公司將寄出替換用產品。

04. This (   ) to the Norelco 6945XL.
    這份保固證明適用於 Norelco 6945XL。

Warranty Status
SD Inkjet 5460 Printer
Your SD warranty expires on Jun 4, 20××

Serial Number: 4DC9403M2D
Product Number: G4809AT
Product Name: SD Inkjet 5460 Printer
Date of Warranty Check: Sep 13, 20××

Warranty Type: Base Warranty
Service Type: SD Maintenance Offsite Support
Status: (Active) / Expired
Start Date: Jun 5, 20××
End Date: Jun 4, 20××
Service Level: Global Coverage, Standard Material Handling, Standard Parts, Customer Visit and Pickup at Repair Center

Product Replacement Instructions

To replace your defective product, you will need to send the package to a nearby service center using a traceable carrier such as UPS or FedEx. We will ship a replacement product after we receive the defective product.

Review our warranty policy.

單字

- warranty status　保固狀態
- warranty type　保固類型
- maintenance　維護
- active　有效
- expired　到期
- standard　標準
- material　材料
- repair center　維修中心
- defective product　瑕疵產品
- traceable carrier　可追蹤的貨運公司
- ship　運送、輸送
- replacement product　替換商品

文書範本 1 公司重整備忘錄 ▶ 翻譯請參閱 p.230

To: All Office Mart associates
From: Nate Kendall, President and Chief Executive Officer
Date: July 11, 20××
Subject: Company Restructuring

Today we are announcing structure changes that will impact many of our dear Office Mart associates. While they were very difficult decisions, I believe they are necessary for the future survival and growth of the company.

Due to a shift in today's market and a negative economic downturn, a need has emerged to readily adapt for future times. We are positioning Office Mart to be a more agile company that readily meets customer demands in the face of stiff competition. As a result of these attempts, 350 associates will be displaced within the next two weeks.

We thank these associates for their service for the company. Let me assure that we will do the best we can to assist them in this difficult period of transition.

You will be hearing from your leadership soon about structural changes occurring in your area. If you have any questions, please communicate with your leaders. Thank you for your understanding in these difficult times.

單字

- company restructuring 公司組織變動
- survival 生存
- economic downturn 經濟衰退
- adapt 適應
- position 安置
- agile 敏捷的、靈活的
- stiff competition 激烈的競爭
- associate （事業、職場上的）同事、員工
- displace 迫使…離開原位
- assure 保證、保障
- transition 轉換
- structural change 結構變化

美國百大企業文書撰寫祕訣 ！

Memo（備忘錄，Memorandum）是公司經常用於內部溝通的文書。主要多由管理階層寫給團隊成員，可以透過電子郵件、更新網站內容或手寫直接交予等方式傳達。備忘錄的內容包羅萬象，包括政策變動、人事異動、現況更新、提出問題，而備忘錄的目的在於使收到備忘錄的人做出適當的措施。

實務負責人的一句話 ！

備忘錄是聯繫撰寫者的目的，和收件人的興趣及需求最有效的橋樑。因此寫作時要掌握收信人的情況，撰寫出相應的備忘錄並確認無誤後再發送。

文書架構詳細解析

①
To: All Office Mart associates
From: Nate Kendall, President and Chief Executive Officer
Date: July 11, 20××
Subject: Company Restructuring

②
Today we are announcing structure changes that will impact many of our dear Office Mart associates. While they were very difficult decisions, I believe they are necessary for the future survival and growth of the company.

③
Due to a shift in today's market and a negative economic downturn, a need has emerged to readily adapt for future times. We are positioning Office Mart to be a more agile company that readily meets customer demands in the face of stiff competition. As a result of these attempts, 350 associates will be displaced within the next two weeks.

We thank these associates for their service for the company. Let me assure that we will do the best we can to assist them in this difficult period of transition.

④
You will be hearing from your leadership soon about structural changes occurring in your area. If you have any questions, please communicate with your leaders. Thank you for your understanding in these difficult times.

❶ 標題（Heading）
寫出收件人和發件人姓名，以及備忘錄的撰寫日期與標題。

❷ 前文（Opening）
提出備忘錄的目的，以及欲於備忘錄中探討的事例及問題。

❸ 本文（Main Body）
詳細記載欲探討的事例或問題。此處也可以寫事例或問題的相關發現（finding）及足以佐證該發現的事實（supporting fact）、建議事項（recommendation）等。

❹ 結語（Closing）
鄭重且詳細地寫出希望收件人採取的應對措施。若有附件資料可在此處提及。

撰寫文書所需的主要用語

01 **be necessary for/to...**　為了 / 對⋯是必要的
This reorganization **is necessary to** maximize profits.
為了獲得最大利益，公司整頓是必要的。

02 **position... to be more...**　使⋯能夠更加⋯
This **positions** our department **to be more** cost-efficient.
這將使本部門的成本使用更有效率。

03 **in the face of stiff competition**　面對激烈的競爭
In the face of stiff competition, we must change and adapt.
面對激烈競爭，我們必須改變和適應。

04 **do the best we can to...**　為⋯盡力做到最好
Kingmaker will **do the best it can to** salvage the situation.
為了挽救局面，造王者公司將全力以赴。

05 **influx of new recruits**　湧入新進員工
The **influx of new recruits** has necessitated the purchase of additional equipment.
因新進員工湧入，有必要添購設備。

06 **foster teamwork and cooperation**　發展團隊合作
This company outing should **foster** more **teamwork and cooperation**.
這次公司出遊應會促進團隊合作精神。

07 **raise employee morale**　提振員工士氣
Company executives must think of ways to **raise employee morale**.
公司主管必須想辦法提振員工士氣。

08 **appreciate feedback and comments**　感謝回饋及評論
We **appreciate** your **feedback and comments** on the new model.
感謝您對新商品的回饋和評論。

小測驗

01. We would (　　　　　　　　　　) and comments on the new office.
　　感謝您對新辦公室的回饋和評論。

02. The company luncheon should (　　　　　　　　).
　　公司午餐宴應會提振員工士氣。

03. The reorganization (　　　　　　　　　) for the survival of the company.
　　為了讓公司存活，公司整頓是必要的。

04. We will do (　　　　　　　　) to remedy the situation.
　　我們將全力以赴去改進這個狀況。

TO: Christopher Sullivan, Roger Brock, Hildred Cho
CC: Keith Jackson, Javier Williams
FROM: Jeana Scott
DATE: March 11, 20××
RE: Training Seminar on May 15

Due to an influx of new recruits, the department as a whole has decided to hold a training seminar at the Grand Hotel on May 15, 20××.

The seminar will cover topics such as company rules, general policies, dress code and business etiquette. Motivational speakers are scheduled to appear and group activities that will foster teamwork and cooperation are being planned. I'm hoping that this seminar will turn into an annual event as I believe that it will greatly raise employee morale.

Please see attached agenda for the training seminar. I would appreciate any feedback and comments.

Attached: May 15 Training Seminar Agenda

單字

- influx 流入、匯集
- department 部門
- training seminar 培訓研討會
- company rules 公司規定
- policy 政策
- dress code 服裝規定

- motivational speaker 勵志演講者
- foster 培養、促進
- cooperation 合作
- employee morale 員工士氣
- attach 附加
- agenda 議程

THE USA

PART 4
Emails

電子郵件

商務郵件的基本原則

大部分上班族都用電子郵件進行必要的工作交流。根據統計,商務郵件(一天 890 億)往來的次數遠高於個人郵件(一天 550 億封)。可見電子郵件是受到無數人青睞的職場溝通方式,也因為被活用於職場上,所以必須寫得更正確詳細。和個人郵件不同,撰寫商務郵件時必須依據一定的格式,遵守基本原則。

1. email 還是 e-mail ?

用英文表示電子郵件時可以用 email 或 e-mail,兩種都可以。E 是 electronic(電子)的縮寫,再加上 mail 後所形成的複合名詞,原則上中間要以連字號連接。其他的例子還有 e-commerce(電子商務)、e-learning(數位學習)、e-business(電子商業)等。不過大部分的使用者、出版社、字典,使用 email 的頻率較高。Co-work 漸漸變為 cowork 也是類似的例子。重要的是,擇一使用以保持一致性。

2. 電子信箱地址怎麼設定 ?

有時會遇到口頭告知對方電子信箱地址,或是對方直接輸入電子信箱地址的情況,所以比起使用複雜的筆名和數字,商務電子信箱地址建議加入本人的名字,越簡單越好。如果一定要加上數字,應避免使用 O、0、1、I 等容易看錯的數字與字母。雖然大部分的電子信箱服務不分大小寫,但是在留 email 給別人時大小文字混用可能會使對方感到困惑,因此要以小寫填寫。

allintheillusion4829118@mycompany.com (X)

KevinPark@mycompany.com (X)

kevinpark@mycompany.com (O)

依「名 - 姓」順序是商務電子信箱地址最常見的設定方式,有必要設計得不同時,請採取下列方式。

- first name + last name = kevinpark
- first name . last name = kevin.park
- first name – last name = kevin-park
- first name + middle initial + last name = kevinepark
- first initial + middle initial + last name = kepark
- first initial + last name = kpark

寄電子郵件的時候，應讓收件人在收件匣能一眼看見寄件人的全名（名和姓）。只出現公司名稱或其他帳號的話，很容易被視為垃圾信。

MyCompany <kevinpark@mycompany.com> (X)

kevinpark@mycompany.com <kevinpark@mycompany.com> (X)

Kevin <kevinpark@mycompany.com> (X)

Kevin Park <kevinpark@mycompany.com> (O)

3. 不用 Dear，改用 Hi 或 Hello 可以嗎？

隨著電子郵件發展為大眾化的溝通手段，商務電子郵件上，也越來越多人使用較親切自然的 Hi/Hello 來取代 Dear。不過第一次寫信給對方，或是寄信給 CEO 等高層人員時，建議使用 Dear。此外，根據彼此關係或親疏程度，也可以選擇用 Hi、Hello、Hey、Hi there、Good morning 等。但無論選用哪一種方式，最好都要加入問候語。下表整理了正式及非正式的問候方式。

問候語	說明
Dear Mr. Smith:	Dear 加冒號（:）是非常正式的用法，在寄信給地位特別高的人或第一次寄重要信件時使用。
Dear Mark,	有人認為此為舊式用法，比較生硬，但也有人認為是基本禮儀，仍持續使用。若不清楚該用哪一種問候方式，此為最保險的選擇。
Hello (Mark),	沒有 Dear 那麼生硬，又比 Hi 更正式一點。是企業寄信給客戶時常見的用法。
Hi (Mark),	熟識的關係間最常用的用法。可以加上人名寫為 Hi Mark, 也可不加名，直接說 Hi 就好。
Hey (Mark),	比 Hi 還更隨興的用法，不過也有人認為不夠體面。
Mark,	有存在不加問候語直接寄信的情況，但會令人覺得語氣不近人情。
「省略」	省略問候語會給人一種沒有誠意的態度，建議不要在撰寫商務電子郵件時省略問候語。
Hi there. / Hello there. / Hey there.	通常用於寫非正式的電子郵件，但不知道對方名字的情況下使用。

4. 什麼時候該用 Mr.、Mrs. 和 Ms.？

第一次寄信就直接叫名字的話，會給人失禮的印象，所以最好遵照 Dear Mr. Jones 這樣的格式。熟識後可以只加上名字，例如 Dear Danny。Mrs. 是已婚女性，Miss 是未婚女性，不知道是否已婚時，則可以用 Ms. 通稱。也可以不用 Mr./Ms.，直接稱呼全名，不區分對方性別（例：Dear Jamie Anderson）。英式英文裡 Mr/Ms 後面不加縮寫點（例：Dear Mr Jones、Dear Ms Gibney）。

5. Re:（或 RE:）

「Re:」表示 Reply（回覆）的意思。在大部分的電子郵件平台上，只要按下回覆按鈕，信件主旨欄就會自動出現「Re:」。但也有人認為這代表 Regarding（關於…），是在初次寄信給某人時使用。大原則是要記得「Re:」是 Reply 的縮寫，視情況有時不一定是回覆的意思即可。

6. Fwd:（或 FWD:、Fw:、FW:）

「Fwd:」是代表「轉寄」之意的 Forward 的縮寫。寫在主旨中代表將收到的信件轉發給其他人。

7. Cc:（或 CC:）

「Cc:」是 Carbon copy 的縮寫，意指「副本」，意思是電子郵件收件人之外的寄件對象。電子郵件中的「To:」欄填寫第一級收件人（primary recipient），「Cc:」欄填寫第二級收件人（secondary recipient）。Carbon copy 是指用碳式複寫紙撰寫的副本，而電子郵件中的 Cc 對象因為形同收到郵件的副本（Copy），所以用這個縮寫來表示。

8. Bcc:（或 BCC:）

「Bcc:」是 blind carbon copy 的縮寫，意思是「密件副本」。除了第一級、第二級收件人外，還要寄信給第三級收件人（tertiary recipient）時，便會填寫在這個欄位。因第一級、第二級收件人看不到第三級收件人的電子信箱地址，所以加上 blind（密件）來表示。一般來說，第三級收件人也無法看到除了本人之外的第一級、第二級收件人，或其他第三級收件人的電子信箱地址。經常用於收件人較多，或是聯絡人清單（mailing list）不欲人知的情況。

9. P.S.

「P.S.」是 Postscript 的縮寫，和紙本書信一樣，是電子郵件中常見的用法。用來在郵件的最後問候語之後，加註單句或整個段落。語源為拉丁語的 post scriptum，意指「written after」，也就是「寫完之後」之意。P.S. 多見於私人郵件，不建議用在正式的商務郵件中。第一次附言後要再附言的話，可以用 PPS（post post scriptum）。雖然不常見，不過只要加 P 就可以無限增加附言（例：PPPS、PPPPS 等）。

10. 電子郵件寫作架構

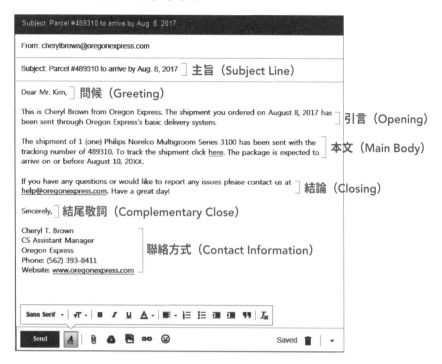

❶主旨（Subject Line）

職場工作者收到信件時，會根據主旨來決定立刻打開來看、之後再看、不用看。因此電子郵件的主旨講求簡潔精確（八個單字以下）。使用手機的話，過長的電子郵件主旨會無法完整顯示。若無法看到完整的主旨，對方會無法辨別信件中的內容，導致不關注信件。近年來有 40% 的電子郵件都是使用手機開啟，因此在撰寫主旨時應力求簡短明瞭，讓人一看就明白郵件概要。若不知道該如何撰寫主旨的話，可用 Regarding 或 About 作為開頭，再加上郵件內容概要，為最不會出錯的做法。

主旨範例	Subject: Parcel #489310 to arrive by Aug. 8, 20××
負面範例	(1) Subject: Here's the IP5 statistics report that you requested from me yesterday. (2) Subject: Meeting (3) Subject: Good morning (4) Subject: To Karen (5) Subject: It's Paul
正面範例	(1) Subject: IP5 statistics report (2) Subject: Meeting request for Tuesday
Regarding, About	[Regarding 或 About（關於）＋ Noun（名詞）] Subject: Regarding our meeting yesterday Subject: About the statistics report

❷問候（Greeting）

電子郵件的第一句話用 Dear、Hi、Hello 作為問候。如果首次寄件的對象職位較高，建議加上姓氏（例：Dear Mr. Price,）。若不知道收件人的名字，可以寫職稱（例：Dear HR Manager,）。To whom it may concern, 和 Dear Sir or Madam, 是舊式用法，有被視為失禮的可能性，盡量不要使用。正式的電子郵件使用冒號（:），一般用逗號（,）即可。若是熟識的對象，也可以直接加名字當開頭，如：Dear Steve,。

> Dear Mr. Kim,

❸引言（Opening）

撰寫電子郵件時，第一段必須寫明寄信的要務（目的）。若是第一次寄信給對方，最好做簡單的自我介紹。如果把目的寫在其他段落，會無法一目了然，所以必須在第一段就使對方掌握寄信用意。

> This is Cheryl Brown from Oregon Express. The shipment you ordered on August 8, 20✕✕ has been sent through Oregon Express's basic delivery system.

❹本文（Main Body）

在本文中以相關事實及附加資訊為基礎詳加描寫，以利具體說明，進一步擴展引言提到的主題。

> The shipment of 1 (one) Philips Norelco Multigroom Series 3100 has been sent with the tracking number of 489310. To track the shipment click here. The package is expected to arrive on or before August 10, 20✕✕.

❺結論（Closing）

在最後一段說明希望對方採取的行動，或向對方致謝。也可以留下自己（寄件人）的聯絡方式作為信件結尾。

> If you have any questions or would like to report any issues please contact us at help@oregonexpress.com. Have a great day!

❻結尾敬詞（Complementary Close）

可放在電子郵件最後的敬詞有很多種，寫信時可能會分不清楚。表格統整從正式到輕鬆等不同風格的敬詞。寄件時只要根據對方的身分、寄件情況挑選適合的敬詞即可。列出的敬詞中以 Best regards, 和 Sincerely, 最常見，不知道要用哪一種時，可以從這兩者中擇一，絕對不會出錯。另外，商務郵件中最好避免「省略」和非常不正式的用法。

最正式	[上司、政府機構、初次聯絡的客戶等] - Respectfully submitted, / Respectfully, / Very truly yours, / Yours truly, / Yours sincerely, / Sincerely yours,
正式	[已經聯絡過幾次的公司職員、客戶] - Sincerely, / Cordially, / Yours faithfully, / Best regards, / Kind regards, / Regards,
較不正式	[有一定交情的職場同事、下屬等] - Best wishes, / With thanks, / Warm wishes, / Warm regards, / All the best, / Many thanks, / Thanks, /「省略」
不正式	[有私交的朋友、熟人] - Take care, / Cheers, / See you soon,
非常不正式	[家人、好友、伴侶] - Much love, / Love, / With love, / Yours forever, / XOXO,

❼聯絡方式（Contact Information）

在電子郵件最後的署名部分留下名字和聯絡方式，這部分可以是 1) 只有名字、2) 放上職稱、公司名稱、電子信箱、公司地址、電話號碼、傳真號碼、公司官網網址、個人社群媒體網址等。一般來說，至少要有姓名、職稱和公司名稱。如果須要收到對方的回覆，最好一併附上電話號碼。依姓名 - 職稱 - 公司名稱 - 聯絡方式的順序撰寫即可。如果對方是熟人，省略這部分也無妨。

Sincerely,

Cheryl T. Brown
CS Assistant Manager
Oregon Express
Phone: (562) 393-8411
Website: www.oregonexpress.com

11. 撰寫商務電子郵件的 12 項基本原則

❶所有段落皆從頁面最左端開始寫，不縮排。

❷段落之間空一行。

❸在第一段開門見山說出要務（目的）。

❹本文盡可能詳細撰寫主旨相關內容。

❺在結論中提到要求對方做的事。

❻活用簡短的單字、句子，力求信件簡明易懂。

❼使用主動語態完成寫作。

❽撰寫信件後仔細檢查是否有錯誤或錯字。

❾訂定簡短又能點出核心內容的主旨。

❿有需要羅列的事項時，使用項目符號（bullet point）。

⓫不使用簡稱。

⓬收到電子郵件後盡快回覆。

文書範本 1 新網站 宣傳廣告信　　　▶翻譯請參閱 p.232

Subject ✉	James Sutton: You're Invited to Join EZ's Site, WearPlus.com

Dear James Sutton,

We'd like to recommend EZ's private shopping site,

WearPlus.com.

We offer great benefits for your shopping pleasure.

On Sale Every Day: Up to 40% off designer brands, plus great finds for home.

Free Delivery: Free shipping on orders over $69

Powered by EZ: Sign in and make purchases
with your existing EZ account.

New event launch daily. Join today!

Unsubscribe | Help

Customer Assistance 1-866-235-5443

©2017 EZ Inc., 1555 Clark Street
Westbury, NY 11590. All rights reserved.

單字

- designer brand 名牌
- private 私人的、私有的
- benefit 利益
- pleasure 愉快
- great find 好物
- sign in 登入
- make a purchase 購買
- existing 既有的、原有的
- unsubscribe 取消訂閱
- access 使用權
- customer assistance 顧客支援
- all rights reserved 版權所有

美國百大企業文書撰寫祕訣！

Promotional Email（宣傳廣告信）是企業基於行銷目的，統一發送給客戶或會員的電子郵件。宣傳廣告信的目的除了通知收件人折扣活動、新產品上市、新服務等消息之外，也可提升品牌知名度，建立與顧客的良好關係。不用花太多成本即可接觸大量用戶，是一種十分有效率的手段，因此有越來越多企業採取這種方式。宣傳廣告信要盡可能減少文字，活用各種醒目的圖片，吸引讀者視覺上的注意。

實務負責人的一句話！

慎選主旨，以免被收件人當作垃圾郵件而忽視。主旨應包含收件人的名字、收件人關注的事情、引人注目的關鍵字等，盡可能提高收件人閱讀的可能性。

文書架構詳細解析

❶ Subject ✉ James Sutton: You're Invited to Join EZ's Site, WearPlus.com

❷ Dear James Sutton,

We'd like to recommend EZ's private shopping site,

WearPlus.com.

We offer great benefits for your shopping pleasure.

❸ On Sale Every Day: Up to 40% off designer brands, plus great finds for home.

Free Delivery: Free shipping on orders over $69

Powered by EZ: Sign in and make purchases
with your existing EZ account.

New event launch daily. Join today!

❹ Unsubscribe | Help

Customer Assistance 1-866-235-5443

©2017 EZ Inc., 1555 Clark Street
Westbury, NY 11590. All rights reserved.

❶ **主旨（Subject）**
主旨要盡量有創意，讓人想展信閱讀。

❷ **標題（Title）**
寫出能明確反映電子郵件內容的標題或介紹文，要簡單扼要、切中核心。

❸ **本文（Body）**
詳細撰寫與宣傳廣告信主題相關的內容，像是折扣活動等，或以產品相關資訊為描寫重點以利宣傳。撰寫本文時最好活用各種圖片來抓住讀者的注意力。

❹ **其他（Other）**
提供取消訂閱連結及公司聯絡方式、資訊等。

撰寫文書所需的主要用語

01 **you're invited to...** 邀請您到…
You're invited to visit our newest location at Ulsan.
歡迎蒞臨本公司在厄爾森新開幕的門市。

02 **we'd like to recommend...** 我們想推薦…
We'd like to recommend our new Android phone, NK1.
我們想向您推薦新款安卓手機 NK1。

03 **up to... % off** …折起
Up to 25% **off**.
75 折起。

04 **free shipping on orders over...** 滿…免運
We offer **free shipping on orders over** $100.
訂單滿 100 美元以上提供免運費服務。

05 **take advantage of benefits** 享受優惠
Take advantage of all these discount **benefits**.
請盡情享受所有的折扣優惠。

06 **get access to...** 取得…、接近…
Get access to these incredible savings by scanning the QR code below.
請掃描下方 QR code 以取得驚人的折扣優惠。

07 **follow us online** 請追蹤我們
Follow us online by registering here.
請加入以追蹤我們。

08 **unsubscribe from future emails** 取消訂閱信件
Unsubscribe from future emails by clicking here.
請點擊此處取消訂閱。

小測驗

01. Get (   ) free MP3 files.
取得免費的 MP3 檔案。

02. We'd like (   ) our newest Internet service.
我們想向您推薦我們最新的網路服務。

03. (   ) at Facebook.
請在臉書追蹤我們。

04. (   ) of many benefits.
享受許多優惠。

| Subject ✉ | Save More than 30% at Headway Plus |

View as a web page | Forward to a friend

Welcome to Headway Plus
Exclusive benefits Terrific Savings

Start taking advantage of your benefits.

ONLINE WEBSITE
Buy on our website and save more than 30%!
Sign Up >

MOBILE APP
Get access to great discount coupons and much more.
Download Now >

OFFLINE STORE
Meet managers who will find the perfect items for you.
Find a Location >

Follow us online.

Please do not reply to this email. To contact us, click here or call us at 1-800-776-5000.

To customize the settings for your profile and subscriptions, click here. To unsubscribe from future emails, click here.

單字

- view 看、閱覽
- forward to... 轉達給…
- exclusive 專屬的、獨占的
- terrific 很棒的
- savings 節省
- sign up 註冊

- take advantage of... 利用…
- get access 接近
- discount coupon 優惠券
- follow online 線上追蹤
- customize 客製
- subscription 訂閱

文書範本 1 請求提供樣本

▶翻譯請參閱 **p.234**

| Subject ✉ | Request for "Active Liver" Sample |

Dear Mr. Spink:

My name is Hyun-jung Lee and I'm in charge of marketing at Han Pharmaceutical. We are the second largest pharmacy chain here in Korea and are constantly looking to expand our product lineup.

We are interested in importing some of your products. We are especially interested in "Active Liver" which I understand promotes healthy liver.

Could you send us a free sample of Active Liver? One unit of the product or 30 tablets should be sufficient. We would be grateful if you could also provide details such as ingredients, recommended use and supplement facts. Please also include pricing information for bulk orders. If lucrative, we are hoping to make an exclusive contract with you.

You may reach me via this email address or by calling me at 82-2-990-3933. I look forward to hearing from you.

Kind regards,

Hyun-jung Lee
Marketing Manager
Han Pharmaceutical

單字

- pharmaceutical 製藥的
- chain 連鎖（店）
- be looking to... 尋找…的方法
- expand 擴張
- product lineup 產品線
- import 進口
- sufficient 充分的
- supplement facts 營養資訊、補充資訊
- bulk order 大量訂購、大宗訂單
- lucrative 賺錢的、有利益的
- exclusive contract 獨占合約
- via... 藉由…、透過…

美國百大企業文書撰寫祕訣！

Request email（請求信）是用來要求對方提供產品樣品或檔案、服務等的信件。撰寫重點在於正確寫出要求提供者的姓名、職責、聯絡方式。若是第一次寄信給對方，應先自我介紹再提出要求。此外，也必須說明提出要求的目的，並在最後一段提供聯絡方式，鄭重表示期待收到對方的回覆。

實務負責人的一句話！

寫請求信的時候，一併把自己的請求為何重要，以及若對方達成請求會有什麼好處寫清楚，才會大大提高收到對方回應的機率。

文書架構詳細解析

❶ Subject ✉ Request for "Active Liver" Sample

❷ Dear Mr. Spink:

My name is Hyun-jung Lee and I'm in charge of marketing at Han Pharmaceutical. We are the second largest pharmacy chain here in Korea and are constantly looking to expand our product lineup.

❸ We are interested in importing some of your products. We are especially interested in "Active Liver" which I understand promotes healthy liver.

Could you send us a free sample of Active Liver? One unit of the product or 30 tablets should be sufficient. We would be grateful if you could also provide details such as ingredients, recommended use and supplement facts. Please also include pricing information for bulk orders. If lucrative, we are hoping to make an exclusive contract with you.

❹ You may reach me via this email address or by calling me at 82-2-990-3933. I look forward to hearing from you.

Kind regards,

Hyun-jung Lee
Marketing Manager
Han Pharmaceutical

❶ **主旨（Subject）**
撰寫主旨時，應寫明「Request for」以表示需要對方提供的產品。

❷ **引言（Opening）**
簡單介紹自己及公司。自我介紹時，除了姓名之外也可以提及職稱。

❸ **本文（Main Body）**
寫清楚自己要向對方提出的請求。告訴對方如果接受並同意請求的話，能夠得到什麼利益或好處。

❹ **結尾（Closing）**
留下聯絡方式並表示期待收到對方的回音，鄭重地為信件作結，並採用 Kind regards, 等態度恭謹的結尾敬語。

撰寫文書所需的主要用語

01 **I'm in charge of...** 我負責⋯
I'm in charge of PR at my company.
我負責本公司的公關工作。

02 **we are interested in...** 我們對⋯有興趣、我們想⋯
We are interested in receiving an estimate on the X1.
我們希望可以收到 X1 的估價單。

03 **Could you send us... ?** 可以請您提供⋯給我們嗎？
Could you send us a brochure on all your products?
可以請您提供貴公司全產品的宣傳資料嗎？

04 **we would be grateful if...** 如果⋯我們會很感謝
We would be grateful if you could also send pricing information.
如果您也可以提供報價資訊的話，我們會非常感激。

05 **you may reach me via...** 可以透過⋯聯絡我
You may reach me via cellphone or email.
您可以透過手機或電子郵件聯絡我。

06 **look forward to...** 期待⋯
I **look forward to** hearing from you.
期待收到您的回覆。

07 **I am writing to...** 來信（聯絡您）是為了⋯
I am writing to request a copy of the MOU.
來信想請您提供合作備忘錄。

08 **thank you for...** 謝謝您⋯
Thank you for your consideration.
謝謝您的諒解。

小測驗

01. Could you (   ) a mustard sample?
可以請您提供一個芥末色樣品給我們嗎？

02. I (   ) to your quick reply.
期盼您迅速回覆。

03. I'm (   ) of marketing and sales.
我負責行銷業務。

04. We would (   ) if you could send an estimate.
如果您可以提供報價單的話，我們會非常感激。

Subject ✉	Recommendation Letter Request

Dear Mr. Berman,

I hope you are doing well. I am writing to ask you a huge favor. I'm applying for a senior marketing position with Kaellum and I was hoping you would write a letter of recommendation on my behalf. Having worked closely with you for more than ten years you were the first person that I thought up for a reference. You also have much credibility and I always respected your judgment and opinions.

For your convenience, I have attached a draft letter which you might use as a template. I included a list of what I think are my core achievements and strengths. I believe they will refresh your memory on the projects that I did between 20×× and 20××.

I completely understand if you don't have the time or inclination to write the letter. If you are willing, feel free to make changes to the attached sample as you see fit. I've also attached my resume and copy of the job posting for your reference. If you have any questions or need more information I can be reached at 324-339-4554.

Thank you for your consideration.

Sincerely,

Amy Leo

單字

- favor 拜託
- on one's behalf 代表某人
- reference 推薦書
- credibility 信賴度
- judgment 判斷
- convenience 便利、方便
- draft 草稿
- core achievement 主要績效
- inclination 意向、傾向
- willing 樂意
- reference 參考
- consideration 考慮、斟酌、關心

提案計畫信

文書範本 1 合作提案

▶ 翻譯請參閱 p.236

Subject ✉ Business Proposal for "Active Liver"

Dear Mr. Spink,

This is Hyun-jung Lee from Han Pharmaceutical. I would like to thank you for the sample of "Active Liver" that we requested from you. It arrived in good condition about a week ago.

After studying the sample and conducting extensive market research we believe that "Active Liver" has good potential here in the Korean market. That's why I propose an exclusive import contract. If you give us exclusive rights for two years to import and sell your product we'll offer the following benefits.

• Wide distribution and promotion of "Active Liver"
• Guarantee of the sale of at least 1,500 units per year
• Future consideration of the importing of other products

Let me know what you think. You can email me or call me at 82-2-990-3933 with your opinion.

Best wishes,

Hyun-jung Lee
Marketing Manager
Han Pharmaceutical

單字

• good condition 良好狀態
• study 研究
• extensive 廣闊的
• market research 市場調查
• potential 潛力
• propose 提案

• import contract 進口合約
• wide 寬廣的
• distribution 流通、分配
• promotion 宣傳
• guarantee 保證
• consideration 考慮、斟酌、關心

美國百大企業文書撰寫祕訣！

Proposal Email（提案計畫信）是用來提議發展各種事業、締結合作關係、簽約的信件。因為目的是要讓對方接受提案，所以要包含能夠說服對方的重點。提案計畫書必須說明對方為何應該接受提案，還有對方接受提案的話有什麼好處。最好在引言就確切提出提案內容以及提案背景。

實務負責人的一句話！

在提案計畫信中，把提案內容和對方關心的議題做連結，得到回覆的機率會更高，因此寫作提案計畫信時，一定要充分考量對方的觀點。

文書架構詳細解析

❶ Subject | Business Proposal for "Active Liver"

❷ Dear Mr. Spink,

This is Hyun-jung Lee from Han Pharmaceutical. I would like to thank you for the sample of "Active Liver" that we requested from you. It arrived in good condition about a week ago.

❸ After studying the sample and conducting extensive market research we believe that "Active Liver" has good potential here in the Korean market. That's why I propose an exclusive import contract. If you give us exclusive rights for two years to import and sell your product we'll offer the following benefits.

• Wide distribution and promotion of "Active Liver"
• Guarantee of the sale of at least 1,500 units per year
• Future consideration of the importing of other products

❹ Let me know what you think. You can email me or call me at 82-2-990-3933 with your opinion.

Best wishes,

Hyun-jung Lee
Marketing Manager
Han Pharmaceutical

---

❶ 主旨（Subject）
要在主旨中提到「Proposal」。

❷ 引言（Opening）
問候對方後，說明這封提案計畫信的寫作背景。

❸ 本文（Main Body）
具體撰寫提案內容，也要提到對方接受提案的話，將得到的回報。如此能提高對方回覆的機率。

❹ 結尾（Closing）
最後請對方回覆對於提案的想法，並留下聯絡方式。

01 I would like to suggest... 我想提議…
I would like to suggest the following.
我想提議的事項如下。

02 conduct market research 進行市場調查
Sam conducted market research for the new model.
山姆為新產品進行市場調查。

03 have good potential 有潛力（可能性）
We believe walnuts have good potential in Vietnam.
我們認為核桃在越南很有市場潛力。

04 That is why I propose... 基於這個原因，我提議…
That is why I propose that we form a partnership.
基於這個原因，我提議建立合作關係。

05 offer the following benefits 提供下列好處
This contract offers the following benefits to your company.
這份合約將提供貴公司下列好處。

06 let me know what you think 請告訴我您的想法
Let me know what you think about my proposal.
請告訴我您對我的提案有什麼想法。

07 require immediate attention 立刻給予關注
This incident requires your immediate attention.
這件事需要您立刻給予關注。

08 I think it would be best if... 我認為如果…的話會是最好的
I think it would be best if you invested in this stock.
我相信投資這支股票，會是您最好的選擇。

小測驗

01. I would (   ) a partnership.
我想提議合作。

02. I think it (   ) if we worked on this together.
我認為和我們合作，會是您最好的選擇。

03. Let me (   ) about my proposal.
請告訴我您對我的提案有什麼想法。

04. I believe the Indonesian market (   ).
我相信印尼市場很有潛力。

| Subject ✉ | Meeting Proposal |

Hi Mike,

This is Lizzy from the HR Department. The recent security leak has not gone well with our customers and the issue requires immediate attention.

This is why I propose a meeting to be attended by us (HR department), the PR Department, Technical Department and the Sales Department. I think it would be best if you attended the meeting yourself as I think you're the most knowledgeable person in Sales.

If it's okay with you, would you let me know when you are available tomorrow?

Lizzy

單字

- recent 最近的
- security leak 安全漏洞
- go well 進行順利
- require 需要
- immediate 即刻
- attention 注意
- attend 參加
- PR Department 公關部門
- Technical Department 技術部門
- Sales Department 業務部門
- knowledgeable 知識淵博的
- available 有空的

詢問信

 文書範本 1 詢問產品　　　　　　　　　　▶翻譯請參閱 p.238

Subject ✉	Inquiry on sparking cables

Dear Ms. Jennings,

My name is John Ha and I'm the production manager at TK Motors. I am looking for a supplier of sparking cables.

I would like to know whether you have them available and, if so, how much they would cost. I'm especially interested to know what types you have and the price for each. Also, I would like to inquire about the extended warranty and payment options.

Please revert back to me with all relevant information and options. Thank you very much.

Best regards,

John Ha
Production Manager
TK Motors
070-3651-9302
www.tkmotors.com

單字

- supplier 供應者、供應商
- sparking cable 火花塞電線
- production manager 產品經理
- cost 花費（金錢、費用等）
- especially 尤其
- be interested in 對⋯有興趣

- type 種類
- inquire 詢問、打聽
- extended warranty 延長保固
- payment option 付款方式
- revert back to... 答覆、回答
- relevant information 相關資訊

美國百大企業文書撰寫祕訣！

Inquiry Email（詢問信）是用來詢問或請求提供所需資訊的信件。大部分的情況下，對方沒有回覆的義務，所以寫作時要以恭謹的語氣詢問，並盡可能引起對方的興趣。寫詢問信的時候先簡單自我介紹，再說明詢問事項及為何詢問，接著提出希望對方提供何種產品資訊。

實務負責人的一句話！

想要讓語氣（tone）鄭重一點的話，比起直接寫問句，建議用「I would like to know...」「I was wondering if...」等委婉的說法來提出要求。

文書架構詳細解析

❶ Subject　Inquiry on sparking cables

❷ Dear Ms. Jennings,

My name is John Ha and I'm the production manager at TK Motors. I am looking for a supplier of sparking cables.

❸ I would like to know whether you have them available and, if so, how much they would cost. I'm especially interested to know what types you have and the price for each. Also, I would like to inquire about the extended warranty and payment options.

Please revert back to me with all relevant information and options. Thank you very much.

Best regards,

❹ John Ha
Production Manager
TK Motors
070-3651-9302
www.tkmotors.com

---

❶ **主旨（Subject）**
比起只寫「Inquiry」，最好以「Inquiry on + 詢問內容」讓對方可以一目了然。

❷ **引言（Opening）**
簡單自我介紹後，說明這封詢問信的寫作背景。

❸ **本文（Main Body）**
具體撰寫詢問內容。建議使用「I would like to know...」等恭謹的語氣，提高對方閱讀的意願。

❹ **結尾（Closing）**
確切說明要拜託對方的事，並以簡單的致謝作結。建議在信件結尾留下聯絡方式，方便對方聯絡。

撰寫文書所需的主要用語

01 **we are looking for...**　我們正在尋找…
We are looking for a methane supplier.
我們正在尋找甲烷供應商。

02 **I'm interested to know...**　我想知道…
I'm interested to know if you give discounts for Ex-Works.
我想知道貴公司與 Ex-Works 交易時是否會提供折扣。

03 **I would like to inquire about...**　我想詢問關於…的問題
I would like to inquire about many of your products.
我想詢問許多關於貴公司產品的問題。

04 **revert back to me**　請回覆我
Please **revert back to me** with more information.
請回覆我進一步的資訊。

05 **come across your company**　偶然發現貴公司
I **came across your company** at LinkedIn.
我偶然在領英上注意到貴公司。

06 **have a strong online presence**　有強大的網路影響力
AT&T **has a strong online presence** nowadays.
AT&T 目前在網路上有著強大的影響力。

07 **I was wondering if...**　我想知道…
I was wondering if you still do business in Korea and Japan.
我想知道您是否還在韓國與日本營業。

08 **explore a possible cooperation**　探尋可能的合作機會
Feel free to contact us if you would like to **explore a possible cooperation**.
如欲洽詢可能的合作機會，歡迎與我們聯絡。

小測驗

01. Please (　　　　　　　　) to me with more details.
　　請回覆我詳細的內容。

02. We have a strong (　　　　　　　　).
　　我們有很大的網路影響力。

03. We (　　　　　　　) for an iron ore supplier.
　　我們正在尋找鐵礦供應商。

04. I'm (　　　　　　　　) about your product lineup.
　　我想知道您的產品線。

120

Subject ✉ Online Services from Collor Ads

Hi there,

My name is Linda Batten and I'm with Collor Ads. We are one of the leading providers of digital marketing services. I came across your organization while searching for companies that might need our services.

You already seem to have a strong online presence and I was wondering if you wanted to further solidify that with our proven online strategies. If you could share with us your goals for this year, I'll gladly explain how we can help you achieve them through online advertising.

If you would like to explore a possible cooperation with us feel free to contact me by email at linda.batten@collor.com or by phone at (353) 477-9004. Looking forward to doing business with you.

Kind regards,

Linda Batten
Collor Ads
www.collor.com

單字

- leading provider 領先供應商
- digital marketing 數位（網路、手機）行銷
- come across 偶然發現
- presence 存在
- solidify 鞏固、固化
- strategy 戰略
- achieve 達成
- online advertising 線上廣告
- explore 探究、探討
- cooperation 合作
- feel free to... 隨時
- do business with... 和…交易

通知信

▶翻譯請參閱 p.240

文書範本 1 服務終止通知

Subject ✉	Shutdown of Netvert Services

Hello Brian Hayden,

We wrote to you in March to let you know that we would be discontinuing the following services:

• Netvert Translator
• Netvert Gallery
• Netvert Hot Jobs

The above services have been discontinued as of August 11, 20×× to streamline our service lineup and to better serve you in the future.

The services have been removed from the main page at www.netvert.com and you can no longer use them. Be assured that we are already working on suitable alternatives.

We apologize for any inconvenience this may cause, and appreciate your assistance and understanding. If you have any questions please visit the Help Center at www.netvert.com/help.

Please do not reply to this email as mail sent to this address cannot be answered.

單字

- discontinue 終止
- translator 翻譯機
- streamline 簡化
- remove 去除、刪除
- no longer... 再也無法⋯
- be assured that... 請對⋯放心
- suitable 適合的、符合的
- alternative 替代方案
- apologize 道歉
- inconvenience 不方便
- appreciate 感謝、歡迎
- assistance 支援、幫助

美國百大企業文書撰寫祕訣！

Notification Email（通知信）是以告知重要資訊為目的的信件。用意和 Memorandum（備忘錄）類似，不過備忘錄是用於公司內部，通知信則是向外部寄送的通知。撰寫通知信時，先提出通知事項，再詳述通知內容及通知原因、參考事項、聯絡方式等。如果希望對方根據通知採取應對措施，可在最後一段提出。

實務負責人的一句話！

通知信是用來寄給顧客或交易對象等外部人士的對外文書，務必以正式的格式寫作。可同時加入其他相關的事項，及對方感興趣的內容。

文書架構詳細解析

① Subject ✉ Shutdown of Netvert Services

②
Hello Brian Hayden,

We wrote to you in March to let you know that we would be discontinuing the following services:

• Netvert Translator
• Netvert Gallery
• Netvert Hot Jobs

③
The above services have been discontinued as of August 11, 20×× to streamline our service lineup and to better serve you in the future.

The services have been removed from the main page at www.netvert.com and you can no longer use them. Be assured that we are already working on suitable alternatives.

④
We apologize for any inconvenience this may cause, and appreciate your assistance and understanding. If you have any questions please visit the Help Center at www.netvert.com/help.

Please do not reply to this email as mail sent to this address cannot be answered.

---

❶ 主旨（Subject）
主旨中放入與通知內容相關的核心關鍵字，讓收件人一目了然。

❷ 引言（Opening）
概述欲通知的內容、核心摘要。

❸ 本文（Main Body）
具體撰寫通知內容，建議再加入通知原因、相關附加資訊以及收件人可能感興趣的事項。

❹ 結尾（Closing）
因應通知內容表示歉意或謝意後，留下聯絡方式作結，以便收件人聯絡。

01 **detect irregular activity**　偵測異常活動
We **detected irregular activity** within your bank transactions.
我們偵測到您的銀行交易明細中有異常活動。

02 **for your protection**　為了保護你
For your protection, we will reset your account's password.
為了保護您的權益，我們會重設您的帳戶密碼。

03 **upon your verification**　您驗證過後
The account will be reactivated **upon your verification**.
帳號將在經過您的驗證後重新啟用。

04 **disregard this notice**　忽略此通知
Please **disregard this notice** if you already paid the subscription.
如您已支付訂閱費用，請忽略此通知。

05 **wrote to you in order to...**　來信給您是為了…
We **wrote to you** last week **in order to** notify you about the price increase.
上週來信給您是為了提醒您漲價一事。

06 **be assured that...**　確定…、對…放心
Please **be assured that** there won't be any more slip ups.
請放心，不會再有任何失誤。

07 **work on suitable alternatives**　籌備合適的替代方案
They are **working on suitable alternatives** to the app.
他們正在規劃適合該應用程式的替代方案。

08 **if you have any questions**　如有任何疑問
If you have any questions, please visit our Help Center.
如有任何疑問請洽客服中心。

小測驗

01. For (　　　　　　　　　　) your password will be reset.
　　為了保護您的權益，您的密碼將被重設。

02. Please (　　　　　　　　　　) that the website will be up soon.
　　請放心，網站即將恢復。

03. If you (　　　　　　　　　　) please call our Help Center.
　　如有任何疑問請電洽客服中心。

04. Please (　　　　　　　　　　) if you don't have an Elastri account.
　　如您沒有埃拉斯特里帳號，請忽略此通知。

Subject ✉ Irregular Credit Card Activity

To: Wendy Parra
Account No. 4474609771
Credit Card No. 8330-1833-1731-9012
Date: 7/30/20✕✕

Dear Valued Customer,

We detected irregular activity on your Bank of Arizona credit card on July 30, 20××. For your protection, you must verify this activity before you can continue using this card.

To review and verify this activity please visit www.bankofarizona.com/protection or calls us immediately at 1-800-832-9412. We will remove any restrictions placed on the account upon your verification. Please disregard this notice if you have already verified the activity.

As this is not a secure form of communication, please do not reply to this email. If you have any questions about your account or need assistance, please call the phone number on your statement or go to "Contact Us" at www.bankofarizona.com.

單字

- **irregular** 不規律的、不正常的
- **valued customer** 重要客戶
- **detect** 探測、發現
- **protection** 保護
- **verify** 證明、證實
- **immediately** 馬上
- **restriction** 受限
- **disregard** 忽視、無視
- **secure** 安全的
- **statement** 對帳單

交易信

▶翻譯請參閱 p.242

文書範本 1 訂單確認信

Subject ✉ Order Confirmation from Wireless.com

Order Confirmation
Order #901-3846119-391

Hello David Lee,

Thank you for shopping with us. Your details for the order you made on October 9 are indicated below. The payment details of your transaction can be found on the order invoice. If you would like to view the status of your order or make any changes to it, please visit "Your Orders" on Wireless.com.

Your estimated delivery date: Tuesday, October 11, 20××

Your shipping type: Standard Shipping

Your order will be sent to: David Lee
4467 Railroad Street
Marquette, MI 49855

Order details:
V-MODA Headphone RS 210
Item Subtotal: $58.95
Shipping & Handling: $2.50
Total: $61.45

Thank you for shopping with Wireless.com. Browse more products and earn store credit here.

單字

- order confirmation 訂單確認
- details 明細
- indicate 標示
- transaction 交易
- invoice 發票
- status 現況、狀態

- estimated 估算的、預計的
- delivery date 交貨日期
- handling 處理
- browse 瀏覽
- subtotal 小計
- store credit 商店紅利

美國百大企業文書撰寫祕訣 ！

Transactional Email（交易信）是客戶在加入服務、購買產品時收到的信件，尤指網路相關交易。舉例來說，加入網站會員收到的歡迎信、密碼設定信、電子信箱驗證信、訂單確認信、發票通知信、延長訂閱申請信等都包含在內。此類信件有很多是自動發送的，不少企業會藉此機會夾帶廣告內容，以達到行銷目的。

實務負責人的一句話 ！

交易信大部分為自動發送，所以無法得到收件人關注的可能性很高。要在主旨和段落中妥善呈現欲傳達的核心資訊，其他附加資訊可以放在末段。

文書架構詳細解析

❶ Subject ✉ Order Confirmation from Wireless.com

❷
Order Confirmation
Order #901-3846119-391

Hello David Lee,
Thank you for shopping with us. Your details for the order you made on October 9 are indicated below. The payment details of your transaction can be found on the order invoice. If you would like to view the status of your order or make any changes to it, please visit "Your Orders" on Wireless.com.

❸
Your estimated delivery date:	Tuesday, October 11, 20××
Your shipping type:	Standard Shipping
Your order will be sent to:	David Lee 4467 Railroad Street Marquette, MI 49855

Order details:
V-MODA Headphone RS 210
Item Subtotal: $58.95
Shipping & Handling: $2.50
Total: $61.45

❹ Thank you for shopping with Wireless.com. Browse more products and earn store credit here.

❶ **主旨（Subject）**
寫明主旨，讓人一眼看出信中提到的交易內容。

❷ **引言（Opening）**
具體說明接下來要討論的交易項目，並明確寫出為何要發送與該交易相關的電子郵件。

❸ **本文（Main Body）**
根據正確資訊，具體提供交易內容。假設是訂單確認信，應要根據訂單成立日期、出貨日期等資訊詳實記載。

❹ **結尾（Closing）**
表示感謝交易後，留下聯絡方式，以便收件人聯絡。另外為了達到行銷效果，可以安插廣告或宣傳內容。

撰寫文書所需的主要用語

01 order details are indicated below 訂單明細如下
The **order details** for your new bicycle **are indicated below**.
您的新款自行車訂單明細如下。

02 can be found on/at... 可在⋯查看
Your receipt **can be found at** your personal page.
您可以在用戶個人頁面查看發票。

03 view the status of your order 查看訂單狀態
To **view the status of your order** check the email we sent you.
如欲確認訂單狀態，請見我們寄給您的電子郵件。

04 estimated delivery date 預計配送日
The **estimated delivery date** of your order is December 1.
您訂購的商品預計將於 12 月 1 日配送。

05 receive a request 收到要求
The department **received a request** to change the design of the website.
部門收到變更網頁設計的要求。

06 click on the link 點擊連結
Please **click on the link** to download the update.
請點擊連結以下載更新版本。

07 link will work for... 連結有效期限至⋯
The below **link will work for** only two hours.
下方連結將在 2 小時後失效。

08 reset your password 重設密碼
The administrator **reset your password** for security purposes.
基於安全目的，管理員重設了您的密碼。

小測驗

01. View the (   ) here.
在這裡查看訂單狀態。

02. Please (   ) the link to download the new app.
請點擊連結以下載這個新的 app。

03. The receipt (   ) on an email we sent you.
您可以在我們寄給您的電子郵件查看發票。

04. To (   ) click here.
請點擊這裡重設您的密碼。

| Subject ✉ | Flex Blogger Password Reset |

Flex Blogger

Greetings Jessica,

We have received a password change request for your Flex Blogger account.

If you made this request, then please click on the link below.

Reset Password

This link will work for 24 hours or until you reset your password.

If you did not ask to change your password, then please ignore this email. Most likely another user entered your username by mistake. No changes will be made to your account unless you click on the link.

The Flex Blogger Team

Follow us on Facebook and Twitter.

單字

- greetings 問候、打招呼
- password 密碼
- request 要求
- account 帳號
- reset 重新設定
- most likely 極有可能會
- enter 輸入
- by mistake 不小心地
- ignore 無視、忽略
- user 使用者
- username 使用者名稱
- click on... 點選…

投訴信

文書範本 1 對有效期限提出抗議

▶翻譯請參閱 p.244

Subject ✉	Expiration date in milk shipment

Dear Mr. Moore,

This is Kara Snyder from Almond Bakery. Yesterday we received the 120 cartons of milk we requested from you. However we found out that the expiration date for 40 cartons had already expired.

We received the package on August 21, 80 arrived in good condition but the rest, 40 cartons in all, had an expiration date of August 19. Since our bakery prides itself in using only the freshest ingredients, we were unable to use the said products. This has caused a delay in production and shortage of our popular mini muffins which resulted in at least 5,000 dollars lost in revenue.

I hope to hear back from you about this unfortunate incident and how you can compensate. Please contact me as soon as possible at (998) 867-0987 or via email at ksnyder@gmail.com.

Respectfully,

Kara Snyder
Almond Bakery

單字

- bakery 烘焙坊、麵包店
- carton 裝飲料等的箱子
- find out that... 發現…
- expiration date 有效期限
- pride itself 為…感到自豪
- fresh ingredient 新鮮的材料
- be unable to... 不可能…
- delay 拖延
- shortage 短缺
- revenue 收益
- unfortunate 不幸的
- compensate 賠償、補償

美國百大企業文書撰寫祕訣！

Complaint Email（投訴信）是使用的服務及產品發生問題或令人不滿時，用來提出抗議的信件。投訴信的送件對象是服務及產品的責任業者（負責人），必須採取鄭重且堅決的語氣撰寫。比起追究錯誤，投訴信的目的更著重於尋找解決對策並獲得賠償，所以最後一段應明確寫出希望對方就發生的問題採取何種處置。

實務負責人的一句話！

投訴信絕對要避免攻擊性的語氣。因可能會讓對方產生被侮辱的感覺，降低對方解決問題的意願。因此要盡量以嚴肅有禮的方式，表達希望對方採取的行動。

文書架構詳細解析

❶ Subject ✉ Expiration date in milk shipment

❷ Dear Mr. Moore,

This is Kara Snyder from Almond Bakery. Yesterday we received the 120 cartons of milk we requested from you. However we found out that the expiration date for 40 cartons had already expired.

❸ We received the package on August 21, 80 arrived in good condition but the rest, 40 cartons in all, had an expiration date of August 19. Since our bakery prides itself in using only the freshest ingredients, we were unable to use the said products. This has caused a delay in production and shortage of our popular mini muffins which resulted in at least 5,000 dollars lost in revenue.

❹ I hope to hear back from you about this unfortunate incident and how you can compensate. Please contact me as soon as possible at (998) 867-0987 or via email at ksnyder@gmail.com.

Respectfully,

Kara Snyder
Almond Bakery

❶ **主旨（Subject）**
使用核心關鍵字撰寫主旨，以便收件人快速掌握信件要投訴的事情。

❷ **引言（Opening）**
明確說明發生的問題為何。假設與產品故障有關的話，要寫出故障產品名稱。如果是與服務有關的問題，則應指明哪一項服務出了問題。

❸ **本文（Main Body）**
具體說明發生什麼問題。盡可能詳述問題是何時、如何發生的，以及因此蒙受了什麼樣的損害。

❹ **結尾（Closing）**
明白說出希望對方針對問題採取何種措施。應避免使用有攻擊性的語氣。

撰寫文書所需的主要用語

01 **in good/bad condition**　處於良好/不良的狀態
The shipment arrived **in bad condition** this morning.
今早送達的貨品狀態不佳。

02 **cause a delay**　導致延遲
The strike **caused a** severe **delay** in the production of the car batteries.
罷工導致車用電池生產嚴重延遲。

03 **I hope to hear back from you**　期待收到您的回覆
I hope to hear back from you soon.
希望盡快收到您的回應。

04 **contact me**　請與我聯絡
Contact me as soon as you get back from your trip.
等您旅遊回來後，請即刻與我聯絡。

05 **be mistakenly given**　被誤給
Amanda **was mistakenly given** the estimate.
阿曼達被誤給了估價單。

06 **put someone under a lot of stress...**　讓…處於壓力之中
The delay in delivery **put me under a lot of stress**.
延遲出貨帶給我相當大的壓力。

07 **put me in a difficult position**　讓我陷入困境
The machine's malfunction **put me in a difficult position**.
機器故障讓我陷入困境。

08 **arrange compensation**　安排補償
Sprint is **arranging compensation** due to the faulty Internet service.
因網路服務錯漏百出，斯畢特公司正在為用戶安排補償。

小測驗

01. This gaffe puts us (   ).
    這個過失讓我們陷入困境。

02. Please (   ) as soon as possible.
    請盡快與我聯絡。

03. The weather (   ) in the delivery.
    因為天氣的緣故導致運送延遲。

04. I was (   ) the wrong package.
    我被誤給了錯誤的包裹。

132

文書範本 2 投訴登機證問題 ▶翻譯及解答請參閱 **p.245**

Subject ✉ Flight No. 121 boarding pass

Dear Nevada Airlines,

I was supposed to board Nevada Airlines Flight No. 121 to Paris on April 21, 20×× when I requested a boarding pass from an airline employee. Unfortunately, I was mistakenly given a boarding pass to another flight.

I didn't realize the error until I had nearly arrived at the wrong gate. This prevented me from arriving at the correct gate in time to make the flight. The error put me in a difficult position. I wasn't able to reschedule a new flight until two days later.

The incident put me under a lot of stress both mentally and financially. I hope your airline can arrange a suitable compensation and action so that this doesn't happen again. I look forward to your prompt response in regard to this matter.

Sincerely,

Christina Griffin

單字

- board 搭乘
- unfortunately 不幸地
- mistakenly 失誤地、錯誤地
- boarding pass 登機證
- gate 大門、登機口
- prevent 防止
- reschedule 重新安排時間、改期
- mentally 精神上
- financially 財務上
- suitable 合適的
- compensation 賠償、補償
- prompt 即時的

邀請信

▶ 翻譯請參閱 p.246

文書範本 1 邀請參加股東會

| Subject | Invitation to Annual Shareholders' Meeting |

Dear Lyncast Shareholder:

You are cordially invited to attend the Annual Meeting of Shareholders of Lyncast, Inc. which will be held at the Orlando Convention Center in Orlando, Florida on Tuesday, February 28, 20×× at 11:00 a.m. Eastern Time. You will find detailed directions in an attached file.

The agenda for this year's meeting is as follows:

1. To approve the 20×× Minutes of Annual Meeting of Shareholders held on March 24, 20××.
2. To consider and approve the company's 20×× balance sheet.
3. To appoint an auditor and remuneration.
4. To consider other business matters.

We kindly ask for your presence at the meeting as there are crucial decisions to be made at this year's meeting. If for some reason you cannot attend please send a representative.

We look forward to seeing you.

Yours faithfully,

Barbara G. Rentro
Executive Chief Officer
Lyncast, Inc.

單字

- shareholders' meeting 股東會議
- shareholder 股東
- cordially 真心地、鄭重地
- agenda 議程
- approve 批准
- appoint 任命、委任
- auditor 審計員
- remuneration 報酬
- ask for... 要求
- presence 出席、在場
- crucial 重大的、決定性的
- representative 代表、代理人

美國百大企業文書撰寫祕訣 ！

Invitation mail（邀請信）是用來邀請收件人出席會議、展覽、研討會等場合的信件。基本上只要告知活動日期、地點、主題等資訊，並確認對方有無出席的意願即可。邀請信最終的目的是讓收件人出席活動，所以須強調出席的重要性，並說出必須出席的原因，有技巧地說服對方，此外也須宣傳希望對方出席的活動。

實務負責人的一句話 ！

邀請信要能夠讓收件人產生參加活動的意願，因此最好提起對方的興趣與活動之間的關聯，強調活動的重要性，並點出對方為何應該要參加。

文書架構詳細解析

❶ Subject ✉ Invitation to Annual Shareholders' Meeting

❷ Dear Lyncast Shareholder:

You are cordially invited to attend the Annual Meeting of Shareholders of Lyncast, Inc. which will be held at the Orlando Convention Center in Orlando, Florida on Tuesday, February 28, 20×× at 11:00 a.m. Eastern Time. You will find detailed directions in an attached file.

❸ The agenda for this year's meeting is as follows:

1. To approve the 20×× Minutes of Annual Meeting of Shareholders held on March 24, 20××.
2. To consider and approve the company's 20×× balance sheet.
3. To appoint an auditor and remuneration.
4. To consider other business matters.

❹ We kindly ask for your presence at the meeting as there are crucial decisions to be made at this year's meeting. If for some reason you cannot attend please send a representative.

We look forward to seeing you.

Yours faithfully,

Barbara G. Rentro
Executive Chief Officer
Lyncast, Inc.

❶ 主旨（Subject）
將活動名稱放進主旨中，讓收件人可以一眼看出受邀參加的活動為何。

❷ 引言（Opening）
鄭重邀請對方參加活動，同時載明活動日期、時間、地點。

❸ 本文（Main Body）
具體撰寫活動內容。例如會議就說明要討論的議題，發表會就介紹要推出的品項。

❹ 結尾（Closing）
最後說明對方應該要參加的理由，並以請對方確認是否出席作結。

01 **you are cordially invited to attend...** 敬邀您參加⋯
You are cordially invited to attend this year's exhibition.
敬邀您參加今年的展覽。

02 **will be held in/at...** 活動將於⋯舉辦
The 2nd annual writers' conference **will be held in** San Francisco.
第二屆作家會議將於舊金山舉辦。

03 **find detailed directions in/at...** 在⋯查看詳細說明
You can **find detailed directions** in the attached file.
詳細說明請見附加檔案。

04 **the agenda is as follows** 議題如下
The agenda for the November 12 meeting **is as follows**.
11 月 12 日會議要探討的議題如下。

05 **ask for your presence** 請您（閣下）出席
We **ask for your presence** at the Deluxe Dining Hall in downtown Seoul.
邀請您蒞臨位於首爾市區的豪華宴會廳。

06 **be pleased to...** 很高興⋯
I **am pleased to** invite you to a luncheon at our company's expense.
很高興邀請您參加以本公司經費舉辦的午餐宴。

07 **directions are enclosed** 內含說明
Directions to the location **are enclosed** for your reference.
內含前往該地點的路徑指引，供您參考。

08 **confirm your attendance** 確認您（閣下）是否出席
Please **confirm your attendance** by August 11.
請於 8 月 11 日前確認您是否出席。

小測驗

01. The expo (   ) in Vancouver.
博覽會將於溫哥華舉辦。

02. We (   ) at our company event.
邀請您參加本公司活動。

03. Directions (   ) for your reference.
內含說明，供您參考。

04. Please (   ) by Friday.
請於星期五前確認您是否出席。

| Subject ✉ | Invitation to S&M Presentation |

Dear Mr. Paulson,

Thank you for joining the Small & Medium-sized Business Association. I'm pleased to invite you to a presentation on the 20×× outlook for small and medium sized businesses.

Details are as follows:

Date: Saturday, January 6, 20××
Time: 2:00 to 4:00 P.M.
Where: Hotel Royale, 2nd floor, 4478 Southside Lane, Los Angeles, CA 90071
Speaker: Dr. Julio Yeomans
Subject: 20×× Outlook for Small & Medium-sized Businesses

We are positive that this presentation will give you helpful ideas for your business for the upcoming new year. Directions are enclosed in a separate file. Please reply to this email to confirm your attendance by January 4, 20××. If you have any questions please contact the association at help@smba.org.

Sincerely yours,

Brenna A. Blair
Chairperson
S&M Business Association

單字

- small and medium-sized business 中小企業
- association 協會
- outlook 展望
- speaker 演講者
- subject 主題
- helpful idea 有幫助的點子
- upcoming 即將發生的
- directions 方向
- be enclosed 隨函附上的
- separate file 個別檔案
- confirm 確認、確定
- attendance 參加

問卷調查信

文書範本 1 問卷調查信 ▶翻譯請參閱 p.248

Subject ✉ Headstart Needs Your Feedback!

Customer Survey

Hello Kyle,

We at Headstart are continually looking for ways to improve our products and services for you. As part of our efforts we would appreciate your participation in this survey. Your feedback provides invaluable information that will help us better meet your needs.

This survey should take approximately 10 minutes to complete. As a thank you to all members who complete this survey, Headstart will forward $1.00 to your balance.

To begin the survey, click on the button below.

Start Survey

If you have any questions or technical problems while taking the survey please email us at survey@headstart.com.

Thank you in advance for your participation. We look forward to hearing from you!

單字

- continually 持續不斷地
- effort 努力
- participation 參與
- provide 提供
- invaluable 貴重的
- survey 問卷調查
- approximately 大約、大概
- complete 完成
- forward 轉寄
- balance 結餘
- technical problem 技術問題
- in advance 預先、事前

美國百大企業文書撰寫祕訣 !

Survey Email（問卷調查信）的目的是了解顧客的需求，調查顧客對公司服務及產品的滿意度。編寫問卷調查信時，應清楚說明問卷調查主旨以及進行問卷調查的原因。此外也需加入「參加問卷可獲得折扣」等誘因，以吸引人們參加問卷調查。

實務負責人的一句話 !

不確定問卷調查需要花多少時間的話，收件人不接受調查的機率很高，所以最好連同問卷調查所需時間一併說明。平均來說，人們每分鐘可完成 5 題選擇題，2 題簡答題。

文書架構詳細解析

❶ Subject | Headstart Needs Your Feedback!

Customer Survey

Hello Kyle,

❷ We at Headstart are continually looking for ways to improve our products and services for you. As part of our efforts we would appreciate your participation in this survey. Your feedback provides invaluable information that will help us better meet your needs.

This survey should take approximately 10 minutes to complete. As a thank you to all members who complete this survey, Headstart will forward $1.00 to your balance.

❸ To begin the survey, click on the button below.

Start Survey

If you have any questions or technical problems while taking the survey please email us at survey@headstart.com.

❹ Thank you in advance for your participation. We look forward to hearing from you!

❶ 主旨（Subject）
以能夠呈現問卷調查內容的關鍵字為中心，編寫出簡短明確的主旨以吸引收件者的注意。

❷ 引言（Opening）
提出問卷調查的目的以及展開問卷調查的動機。

❸ 本文（Main Body）
標明問卷調查所需時間以及參與問卷調查的回報，並詳細記載問卷調查相關內容。除了詳細內容之外，也要放上連結（link）以便收件人進入問卷調查頁面。

❹ 結尾（Closing）
留下聯絡方式，讓收件者可以聯絡，並感謝對方參加問卷調查。

01 **as part of our efforts**　作為我們努力的一環
As part of our efforts to bring you better services, we are upgrading our website.
作為我們努力提供更好服務的一環，我們正在升級網站。

02 **meet your needs**　達到您（閣下）的要求
To **meet your needs**, our company is conducting this short survey.
本公司進行這項簡短的調查，是為了滿足您的需求。

03 **as a thank you**　表示對您的感謝
As a thank you to our loyal customers, we are sending a gift card.
我們會發送禮品卡，以表示對忠實客戶的感謝。

04 **thank you in advance**　先謝謝
Thank you in advance for your cooperation.
先謝謝您的配合。

05 **take a short survey**　參加簡短的調查
Please take a minute to **take this short survey**.
請花一點時間參加這份簡短的調查。

06 **express your opinions**　表示您（閣下）的意見
Please use this chance to **express your opinion** on our services.
請把握這次機會，說出您對本公司服務的意見。

07 **click the link below**　請點擊下方連結
Click the link below to take part in the survey.
請點擊下方連結參加問卷調查。

08 **appreciate your input**　謝謝您（閣下）的貢獻
Sysco would **appreciate your input**.
西斯科公司十分感激大家的貢獻。

小測驗

01. Please take a moment to take a (　　　　　　　　　　).
請花一些時間參加這份簡短的調查。

02. Thank you in (　　　　　　　　　) for your cooperation.
先謝謝您的合作。

03. To (　　　　　　　　　) we are conducting this short survey.
我們進行這項簡短的調查是為了滿足您的需求。

04. Please (　　　　　　　　　) on our products and services.
請說出您對本公司的產品和服務的意見。

Subject ✉ Share your opinion about your selling experience.

Clear Market

Dear Amanda Fisher,

Thank you for being a valued Clear Market seller. Our records show that you sold 8 items within the last week at Clear Market. We would like to invite you to take a short survey to express your opinions about your selling experience on Clear Market. Your input will help make Clear Market a better place for our valued buyers and sellers.

Please click the link below to start the survey right away. It will only take 5 minutes! Or copy and paste the following address into your browser.

https://survey.clearmarket.com/survey/seller/clr88301?co+us&smpl_gst=146827313

We appreciate your input and thanks for using Clear Market!

Best regards,

Clear Market Customer Experience Team

單字

- valued 貴重的
- record 紀錄
- item 物品
- express 表達
- opinion 意見、見解
- selling experience 販售經驗
- input 建議、反饋
- buyer 買家
- seller 賣家
- right away 馬上、即刻
- copy and paste 複製貼上
- appreciate 感謝、歡迎

感謝信

 文書範本 1 對主題演講表示感謝

▶ 翻譯請參閱 p.250

Subject ✉	Thanks for the keynote

Hi Craig,

Thanks for taking time from your busy schedule to deliver the keynote speech last Friday at our company. Your chosen topic of the power of strong communication skills is still resonating as we speak.

To say the least your lecture was very educational and entertaining. It motivated and inspired many of our employees to hone their communication skills. Also the images and supplementary material that you prepared helped us understand your message even better. We sincerely appreciate your efforts.

I hope we can work with you again soon. Congratulations on a job well done!

Sandra Colegrove
HR Manager
Calhoun Logistics

單字

- busy schedule 忙碌的行程
- keynote speech 主題演講
- communication skills 溝通能力
- resonate 引發共鳴、發出迴響
- as we speak 與此同時
- educational 教育的

- entertaining 有趣的
- motivate 使產生動機
- inspire 激發靈感
- hone 磨練、訓練
- supplementary material 補充資料
- sincerely 真心地

美國百大企業文書撰寫祕訣！

比起其他類型的信件，職場員工較少寄 thank you email（感謝信）。不過其實寄感謝信的機會很多，例如收到要求的檔案或工作上受到協助的時候，就可以寄感謝信表達謝意，讓自己在對方心中留下有禮貌的好印象，能進一步鞏固商務關係及人際關係。

實務負責人的一句話！

感謝信不只可以留給對方好印象，也會為自己帶來好處。因此就算篇幅簡短也無妨，要養成寄感謝信傳達謝意的習慣。

文書架構詳細解析

❶ Subject ▣ Thanks for the keynote

Hi Craig,

❷ Thanks for taking time from your busy schedule to deliver the keynote speech last Friday at our company. Your chosen topic of the power of strong communication skills is still resonating as we speak.

❸ To say the least your lecture was very educational and entertaining. It motivated and inspired many of our employees to hone their communication skills. Also the images and supplementary material that you prepared helped us understand your message even better. We sincerely appreciate your efforts.

❹ I hope we can work with you again soon. Congratulations on a job well done!

Sandra Colegrove
HR Manager
Calhoun Logistics

❶ **主旨（Subject）**
感謝信的主旨不需要像其他商務郵件一樣有創意或醒目，只要有用 "Thank you for..." 開頭就足夠了。

❷ **引言（Opening）**
確切說明感謝對方的原因，並以鄭重的語氣致謝。

❸ **本文（Main Body）**
具體描述自己覺得感謝的事情。例如對方做了什麼事，這件事對自己有什麼幫助等。

❹ **結尾（Closing）**
最後表示往後也請多多指教，再次道謝為信件作結。

01 **Thanks for taking time from your busy schedule to...** 感謝您百忙之中撥冗…
Thanks for taking time from your busy schedule to visit us.
感謝您百忙之中撥冗到訪。

02 **to say the least** 至少可以說
To say the least, you did the best you could.
至少可以說，你已經盡力了。

03 **appreciate your efforts** 感謝您（閣下）的努力
I **appreciate your efforts** in building rapport with our clients.
感謝您努力與我們的客戶建立融洽關係。

04 **hope we can work with you again** 希望我們能再次與您共事
I **hope we can work with you again** soon.
希望能夠早日再次與您共事。

05 **I very much enjoyed...** 我很享受…、樂在其中…
I very much enjoyed your hospitality.
我很享受您的款待。

06 **feel free to...** 儘管（隨時）…
Feel free to contact me anytime.
歡迎隨時與我聯絡。

07 **thank you once again** 再次向您（閣下）致謝
Thank you once again for sending us the sample we requested.
再次感謝您寄來我們要求的樣品。

08 **extend kindness** 釋出善意
Thanks for the **kindness** you **extended** during our stay.
謝謝您在我們住宿期間親切以待。

小測驗

01. I ( ) in promoting the X17.
感謝您努力推銷 X17。

02. To ( ), I very much enjoyed your hospitality.
至少可以說，我非常享受您的款待。

03. I hope we ( ) again in the future.
希望未來能再次與您共事。

04. Thanks for taking time from ( ) to reply.
感謝您百忙之中撥冗回信。

Subject ✉	Thank You for the Interview

Dear Mr. Green,

I appreciate the time you took to interview me today. I very much enjoyed meeting you and learning more about the engineer position at Mind Tech.

My visit with you made me even more interested in becoming a part of Mind Tech's staff. I feel confident that my experience as an intern at PXO will help me perform the duties of the position effectively.

Please feel free to contact me if I can provide you with any further information. I look forward to hearing from you and thank you once again for the kindness you extended me.

Yours sincerely,

Carolyn Fink

單字

- **appreciate** 感謝、歡迎
- **interview** 面試
- **position** 職位
- **interested in...** 對⋯有興趣
- **I feel confident that...** 我對⋯有信心
- **as...** 作為⋯、身為⋯
- **duty** 義務、責任
- **perform** 進行
- **effectively** 有效地
- **feel free to...** 隨時
- **kindness** 親切、體貼
- **extend** 提供、給予

THE USA

PART 5
Contracts

合約

合約的基本原則

1. 何謂合約？

企業之間在進行重要交易時會簽約。合約中必須明示欲交易的服務、產品，以及交易時合議之事項、交易條件、期間（期限）等。需注意的是必須預先設想雙方在合約關係中，可能會發生的各種問題，且關於各個問題都必須以具體且明確的句子書寫，應避免會產生任何歧義的表述方式。合約的目的是為了留下交易關係的根據、保護企業的事業股份及避免發生糾紛。當產生糾紛時，合約就能成為相關的佐證資料。在美國通用的商業合約種類如下。

〔合約的種類〕

Sales Contract（買賣合約）   Employment Contract（聘僱合約）

Licensing Contract（授權合約）   Non-disclosure Agreement（保密合約）

Lease Agreement（租賃契約）   Memorandum of Understanding（備忘錄）

Letter of Intent（意向書）   Partnership Agreement（合作協議）

2. 商業合約的格式

❶ **CONTRACT FOR THE SALE OF GOODS**

❷ This Sales Contract (herein "Contract") is made effective as of February 14, 20××, by and between Smith Company (herein "Seller") of 641 Bird Spring Lane, Houston, Texas and Benavides Company (herein "Buyer") of 1450 Point Street, Chicago, Illinois.

❸ GOODS PURCHASED. Seller agrees to sell, and Buyer agrees to purchase the following product(s) in accordance with the terms and conditions of this contract.

Description	Quantity	Unit Price	Total Price
Smith TY189 Chipset	200	$55.00	$11,000.00

PAYMENT. Payment shall be made by Buyer to Seller in the amount of $11,000.00 in cash upon delivery of all Goods described in this Contract.

DELIVERY. Seller will arrange for delivery by carrier chosen by Buyer. Delivery shall be duly completed by February 28, 20××.

WARRANTY. Seller warrants that the Goods shall be free of substantive defects in material and workmanship.

❹ This Contract shall be signed on behalf of Smith Company by Ellen Crane, Sales Representative and on behalf of Benavides Company by Larry M. Guillemette, CFO.

Seller: Buyer:
Smith Company   Benavides Company

Ellen Crane   *Larry M. Gillemette*
Ellen Crane   Larry M. Gillemette
Sales Representative   CFO

❶標題

合約最上方要先打上標題，可以簡單地用「Contract」當標題，或是具體地加上商品或產品名稱。

eg. Website Maintenance Contract
Contract for Cleaning, Housekeeping and Janitorial Services

還有經常也可以看到用 Agreement 取代 Contract，這兩個字都可以當作「合約」來使用，沒有特別區分，一般可以混用。此外，按各家公司的習慣，也會在標題的上方或下方標示公司的名稱。

❷前言（當事人及目的）

合約內會寫出 2 名以上締約的當事人。當事人可以寫上公司或機構的名稱，也可以寫上人名。必須明示當事人須履行的合約內容及目的。

❸主文（合約事項）

雖然合約中的各項條款會因合約種類而異，但是大部分會包含下列內容。

- ・當事人的權利
- ・當事人的義務
- ・合約的履行時間
- ・合約的履行方式

- ・合約的期限及續約
- ・款項支付（金額、方式、時間）
- ・保密
- ・違約及解約（發生事由、事後處理、損害賠償）

❹簽名

在雙方認定的代理人簽名後，合約便具有法律效力。簽名就代表當事人理解並接受合約內容。若是電子合約，採電子簽名、掃描簽名等方式簽約，也都視為有效合約，但目前西方還是偏好親筆簽名。

3. 寫商業合約的技巧

❶以好理解的語言書寫

雖然合約有專門的書寫語言，但最重要的還是雙方都能徹底了解合約內容，因此還是盡可能地以簡單的句子書寫。

❷各條款或段落以「標題和編號」標示

合約可能會多至數十張，所以為了好整理，建議各條款或段落都能加上編號和標題。此外，一項條款中只寫與該條款相關的內容，且為了提高可讀性，亦可將各個段落的標題以大寫字母或粗體字（bold）標示。

❸將內容寫得具體又簡潔

為了不讓合約內容有誤會的空間，必須具體寫下當事人的權利與義務。例如「應於每月中旬移交」應該寫成「應於每月 15 日移交」，屏除曖昧不清的說法，使用明確的語句來表達。利潤也應該明確定義，是總利潤還是淨利潤。即使是再瑣碎的事，也須達成書面協議，比較不會發生問題。還有合約越長，越可能寫進非必要的條件及義務，所以要特別注意。此外，合約內容要盡量寫得讓非專業人士或第三方的人也能輕易理解。

❹數字要用文字和阿拉伯數字一起標示

合約中數字很重要，偶爾會發生在寫數字時標錯逗點，或少一個「0」等失誤。因此在寫數字時，為了防止打錯字或讀錯，建議文字和阿拉伯數字一起寫。例如：ten(10)。

❺先預想可能會打官司的情形

在寫合約的任何一項條款時，都要先預想打官司的情形。尤其是重要條款的內容，必須避免模稜兩可的表達。如果有一項條款，產生訴訟的機率很高，就更應該仔細地書寫相關內容，避免產生對自己不利的後果。

❻檢查是否有錯字或多餘的內容

合約是重要的法律文件，所以必須仔細檢查是否有錯字或多餘的內容。

買賣合約

文書範本 1 物品買賣合約

▶翻譯請參閱 p.252

CONTRACT FOR THE SALE OF GOODS

This Sales Contract (herein "Contract") is made effective as of February 14, 20××, by and between Smith Company (herein "Seller") of 641 Bird Spring Lane, Houston, Texas and Benavides Company (herein "Buyer") of 1450 Point Street, Chicago, Illinois.

GOODS PURCHASED. Seller agrees to sell, and Buyer agrees to purchase the following product(s) in accordance with the terms and conditions of this contract.

Description	Quantity	Unit Price	Total Price
Smith TY189 Chipset	200	$55.00	$11,000.00

PAYMENT. Payment shall be made by Buyer to Seller in the amount of $11,000.00 in cash upon delivery of all Goods described in this Contract.

DELIVERY. Seller will arrange for delivery by carrier chosen by Buyer. Delivery shall be duly completed by February 28, 20××.

WARRANTY. Seller warrants that the Goods shall be free of substantive defects in material and workmanship.

This Contract shall be signed on behalf of Smith Company by Ellen Crane, Sales Representative and on behalf of Benavides Company by Larry M. Guillemette, CFO.

Seller: Buyer:
Smith Company   Benavides Company
Ellen Crane   *Larry M. Gillemette*
Ellen Crane   Larry M. Gillemette
Sales Representative   CFO

單字

- made effective 生效
- good 商品、貨物
- unit price 單價
- payment 支付
- delivery 運送
- warranty 保證書、保用單、保固卡
- warrant 承擔、保證
- substantive 實質的
- defect 瑕疵
- workmanship 製作技術（技藝）
- on behalf of... 代替（代表）…
- sales representative 業務代表

美國百大企業文書撰寫祕訣！

Sales contract（買賣合約）是在轉移產品、服務或財產所有權時所製作的文件，通常是賣方根據合約中的條款，向買方收取款項後交出所有權。而買賣合約為將企業間複雜的買賣交易以白紙黑字寫下，做為證據的文件。合約當事人必須清楚理解和遵循合約中的條款。

實務負責人的一句話！

因為買賣合約是合法契約，因此合約中的資訊必須正確。尤其買賣雙方的資訊、交易物品的基本資訊、買賣價格、買賣日期、移交方式等，都務必寫清楚。

文書架構詳細解析

❶ CONTRACT FOR THE SALE OF GOODS

❷ This Sales Contract (herein "Contract") is made effective as of February 14, 20××, by and between Smith Company (herein "Seller") of 641 Bird Spring Lane, Houston, Texas and Benavides Company (herein "Buyer") of 1450 Point Street, Chicago, Illinois.

❸ GOODS PURCHASED. Seller agrees to sell, and Buyer agrees to purchase the following product(s) in accordance with the terms and conditions of this contract.

Description	Quantity	Unit Price	Total Price
Smith TY189 Chipset	200	$55.00	$11,000.00

PAYMENT. Payment shall be made by Buyer to Seller in the amount of $11,000.00 in cash upon delivery of all Goods described in this Contract.

❹ DELIVERY. Seller will arrange for delivery by carrier chosen by Buyer. Delivery shall be duly completed by February 28, 20××.

WARRANTY. Seller warrants that the Goods shall be free of substantive defects in material and workmanship.

This Contract shall be signed on behalf of Smith Company by Ellen Crane, Sales Representative and on behalf of Benavides Company by Larry M. Guillemette, CFO.

❺ Seller:
Smith Company
Ellen Crane
Ellen Crane
Sales Representative

Buyer:
Benavides Company
Larry M. Gillemette
Larry M. Gillemette
CFO

❶ 標題（Title）
以 Sales Contract, Contract for the Sale of... 等句子來下標。

❷ 當事人（Parties）
寫下簽約日期和買賣雙方的公司名及住址。

❸ 物品（Goods）
除了明示交易物品的名稱之外，還要連物品的單價、總額都清楚寫下。在寫價格時，要多次確認，避免發生錯誤。

❹ 條款（Terms）
寫下移交物品的負責人、移交時間、移交方法等，像保固（Warranty）這種其他相關內容也要寫在此。

❺ 簽名（Signature）
合約當事人需簽名以保證合約的效力。

01　be made effective as of...　從（日期）開始生效
This agreement **is made effective as of** January 20, 20××.
本合約自 20×× 年 1 月 20 日起生效。

02　by and between...　…之間
This contract is made **by and between** Sears and Robinson.
此為西爾斯和羅賓森之間簽訂的合約。

03　in accordance with...　根據…
SK Group agrees to acquire the enterprise **in accordance with** this contract's terms.
SK 集團同意依此合約的條件收購企業。

04　arrange for delivery　準備運送
The Seller will **arrange for delivery** of the goods by March 1.
賣方須於 3 月 1 日前準備好運送貨物。

05　be free of defects　沒有瑕疵
All goods are to **be free of defects**.
所有貨物必須沒有瑕疵。

06　on behalf of...　代替…
This contract will be signed **on behalf of** Seller by Jack Johnson.
本合約由傑克‧約翰遜代替賣方簽署合約。

07　the agreement is made on...　本合約簽訂於…
This agreement is made on October 17, 20××.
本合約簽訂於 20×× 年 10 月 17 日。

08　have full right and authority　擁有所有權利
Intel **has full right and authority** to apply for compensation.
英特爾擁有申請賠償的所有權利。

小測驗

01. This (　　　　　　　　　　　) on June 30, 20××.
本合約簽訂於 20×× 年 6 月 30 日。

02. This contract is made (　　　　　　　　　　　) Tronic and Verbrid.
此為創尼克和費爾比德之間簽訂的合約。

03. The Seller will (　　　　　　　　　) of the goods.
賣方將準備運送貨物。

04. This contract is (　　　　　　　　　) as of May 1, 20××.
本合約自 20×× 年 5 月 1 日起生效。

Agreement for Sale of Business

This Agreement is made on May 17, 20××

BETWEEN   SD Corporation (the "Seller"), with its head office located at Keunwoomul-ro 75, Mapo-gu, Seoul, Korea

AND   TK Corporation (the "Buyer), with its head office located at Sajikro2-gil 48, Jongno-gu, Seoul, Korea.

1. Description of Business
The Business includes the following properties: The inventory, raw materials and finished goods to be transferred from Seller to Buyer as under this Agreement.

2. Purchase Price and Method of Payment
The Buyer shall pay the Seller the sum of $14,200,000.00 USD, inclusive of all sales taxes, paid by certified check.

The Seller warrants that (1) the Seller is the legal owner of the Business; and (2) the Seller has full right and authority to sell and transfer the Business.

The Buyer has been given the opportunity to inspect the Business and related Property and the Buyer has accepted the Business in its existing condition. This Agreement will be construed in accordance with and governed by the laws of the Republic of Korea.

Signed, Sealed and Delivered this 17th day of May, 20×× in the presence of

Seller   Buyer
SD Corporation   TK Corporation
Chin-ho Lym   *Young-min Ock*

單字

- head office 總公司
- property 資產
- raw material 原料
- transfer 轉移、轉讓
- inventory 庫存、存貨
- finished good 成品

- inclusive of... 包含…
- right and authority 權利
- legal owner 法定所有人
- inspect 檢查
- construe 把…理解為、把…解釋為
- be governed by... 受制於…

153

文書範本 1 銷售專員聘僱合約 ▶翻譯請參閱 p.254

EMPLOYMENT CONTRACT

This Employment Contract (herein "Contract") is made effective as of February 14, 20××, by and between Rocklyn Corporation (herein "Employer") and Rudy L. Adams (herein "Employee").

1. EMPLOYMENT. Employer hereby employs the Employee as a sales associate for the period beginning March 1, 20×× and ending on the date on which the employment is terminated. Employee agrees to devote fully to the sales affairs of the Employer's products and goods and perform his duties faithfully, industriously and to the best of Employee's ability and experience. Work hours are 40 hours a week.

2. COMPENSATION. As compensation for the services provided by Employee, Employer will pay an annual salary of $30,000 in accordance with payroll procedures.

3. CONFIDENTIALITY. Employee agrees that Employee will not at any time divulge, disclose or communicate any company confidential Information to any third party without the prior written consent of Employer. A violation of this will justify legal and/or equitable action by Employer which may include a claim for losses and damages.

4. BENEFITS. Employee shall be entitled to 21 days of paid vacation and 5 days of sick leave per year.

5. TERM/TERMINATION. This Contract may be terminated by Employer upon 1 month written notice, and by Employee upon 1 month written notice.

Laura Swann
Laura Swann, HR Manager
Rocklyn Corporation (EMPLOYER)

Rudy L. Adams
Rudy L. Adams
(EMPLOYEE)

單字

- terminate 結束、終止
- affairs 工作、業務
- industriously 勤勞地、勤奮地
- annual salary 年薪
- divulge 洩漏
- legal action 法律措施
- written notice 書面通知

美國百大企業文書撰寫祕訣 ！

Employment contract（聘僱契約）是公司和員工針對年薪、福利、上班時間、工作條件等達成協議後，寫下協議內容的文件。聘僱合約適用於正職員工、約聘員工、時薪人員，若為時薪制勞工，合約中提及薪水的部分應標示每小時的薪資。若不另外製作保密協定，那麼可以在合約書中直接加上公司智財權的保密條款。此外合約中還可加入福利、津貼等內容。

實務負責人的一句話 ！

在撰寫聘僱合約前，包含薪資在內的所有工作條件，都必須在事前達成協議，尤其和薪資相關的事項一定要確實寫清楚。離職的執行程序也須寫進合約中。

文書架構詳細解析

EMPLOYMENT CONTRACT

❶ This Employment Contract (herein "Contract") is made effective as of February 14, 20××, by and between Rocklyn Corporation (herein "Employer") and Rudy L. Adams (herein "Employee").

❷ 1. EMPLOYMENT. Employer hereby employs the Employee as a sales associate for the period beginning March 1, 20×× and ending on the date on which the employment is terminated. Employee agrees to devote fully to the sales affairs of the Employer's products and goods and perform his duties faithfully, industriously and to the best of Employee's ability and experience. Work hours are 40 hours a week.

❸ 2. COMPENSATION. As compensation for the services provided by Employee, Employer will pay an annual salary of $30,000 in accordance with payroll procedures.

❹ 3. CONFIDENTIALITY. Employee agrees that Employee will not at any time divulge, disclose or communicate any company confidential Information to any third party without the prior written consent of Employer. A violation of this will justify legal and/or equitable action by Employer which may include a claim for losses and damages.

❺ 4. BENEFITS. Employee shall be entitled to 21 days of paid vacation and 5 days of sick leave per year.

❻ 5. TERM/TERMINATION. This Contract may be terminated by Employer upon 1 month written notice, and by Employee upon 1 month written notice.

Laura Swann   *Rudy L. Adams*

❼ Laura Swann, HR Manager   Rudy L. Adams
Rocklyn Corporation (EMPLOYER)   (EMPLOYEE)

---

❶ 標題和當事人（Title and Parties）
標題為 Employment Contract 或 Employment Agreement，同時寫下雇主和受雇人。

❷ 職務（Position and Duties）
具體寫下受雇人要負責的職務，以及該職務的基本業務為何。

❸ 報酬（Compensation）
這部分為薪資（工資）相關內容。主要寫的是年薪，但若受雇人為時薪制勞工，則寫每小時的薪資。

❹ 保密（Confidentiality）
寫下受雇人必須遵守的保密相關規定。

❺ 福利（Benefits）
寫下和受雇人協議好之保險、員工福利等相關事項。

❻ 期間／終止（Term/Termination）
寫下聘僱期間及聘雇終止條件。

❼ 簽名（Signature）
雇主和受雇人雙方須簽名。

撰寫文書所需的主要用語

01 **hereby employs... as a...** 特此聘用⋯
The Company **hereby employs** Employee **as** a laboratory technician.
公司特此聘用員工為實驗室技術員。

02 **terminate employment** 終止聘僱合約
Employer or Employee may **terminate employment** at any time.
雇主或受雇人可隨時終止聘僱合約。

03 **perform one's duties faithfully** 忠實履行⋯的義務
The Employee is expected to **perform his duties faithfully**.
員工應忠實履行本人之義務。

04 **to the best of one's ability and experience** 盡⋯的能力和經驗行事
The Executive will act **to the best of his ability and experience**.
主管應盡本人的能力和經驗全力以赴。

05 **pay an annual salary of...** 支付相當於⋯的年薪
The Company will **pay an annual salary of** $26,000 to Employee.
公司將支付員工年薪 26,000 美金。

06 **upon one month written notice** 一個月前以書面通知
Employer may terminate employment **upon one month written notice**.
雇主若欲終止聘僱合約，需於一個月前以書面通知。

07 **be entitled to severance** 有權獲得遣散費
Under the terms of employment, the Executive **is entitled to severance**.
根據聘僱條款，主管有權獲得遣散費。

08 **renew the agreement** 更新合約
Both sides are entitled to **renew the agreement**.
雙方有權更新合約。

小測驗

01. The Company will pay an (   ) of 18,000 dollars.
公司將支付年薪 18,000 美金。

02. The Company (   ) Charles Hill as a salesperson.
公司特此聘用查爾斯·希爾為銷售員。

03. The Employee is expected to perform his (   ).
員工應忠實履行本人之義務。

04. Employer may (   ) at any time.
雇主可隨時終止聘雇合約。

Keystone Construction
Employment Agreement

This Agreement is entered into on July 1, 20××, by and between Keystone Construction Co. (hereinafter referred to as the "Company") and John Norris (hereinafter referred to as "Executive").

Duties and Scope of Employment. As of August 1, 20××, Executive will serve as President and Chief Executive Officer of the Company, reporting to the Company's Board of Directors. During the Employment Term, Executive will devote Executive's full business efforts and time to the Company and will use good faith to discharge Executive's obligations under this Agreement to the best of Executive's ability.

Termination. Executive and the Company acknowledge that this employment relationship may be terminated at any time, upon written notice to the other party, with or without good cause, at the option either of the Company or Executive. However, Executive may be entitled to severance and other benefits depending upon the circumstances of Executive's termination of employment.

Term of Agreement. This Agreement will have a term of four (4) years commencing on the Effective Date. No later than ninety (90) days before the end of the term of this Agreement, the Company and Executive will discuss whether and under what circumstances the Agreement will be renewed.

Compensation. As of the Effective Date, and until June 30, 20××, the Company will pay Executive an annual salary of $400,000 as compensation for his services.

單字

- hereinafter 以下、下文
- refer to... 提到、提及
- board of directors 董事會
- devote 奉獻（努力、時間等）
- use good faith 公平且正直地
- discharge obligation 履行義務

- acknowledge 承認、認可
- severance 資遣費、遣散費
- termination 結束、終止
- commence 開始
- effective date 合約生效日期
- renew 延長、更新

授權合約

 文書範本 1 著作利用授權合約 ▶翻譯請參閱 p.256

COPYRIGHT LICENSE AGREEMENT

This LICENSE AGREEMENT (the "Agreement) is made and entered into effective as of May 15, 20×× (the "Effective Date"), by and between ART TECHNOLOGIES (the "Licensor") and MASON CO. (the "Licensee").

Grant of License. In accordance with the terms and conditions of this Agreement, Licensor hereby grants to Licensee a non-exclusive, non-transferable license to use the Work in the course of its business and to otherwise copy, make, use and sell the Work, and for no other purpose. Any other use shall be made by Licensee only upon the receipt of prior written approval from Licensor.

Term and Termination. This Agreement shall commence as of the Effective Date and shall continue in full force and effect for a period of two (2) years, and shall automatically renew for additional one year periods, unless either party provides written notice of non-renewal to the other party, not less than sixty (60) days prior to the expiration of any one (1) year term.

Fees. Licensee shall pay the Licensor a royalty of five percent (5%) of gross receipts from sale of the Work each month. All royalties are to be paid within ten (10) days of the end of each month.

IN WITNESS WHEREOF, the undersigned have executed this Agreement as of the date first above written.

Licensor
ART TECHNOLOGIES

_____

Licensee
MASON CO.

_____

單字

- **non-exclusive** 非專屬、非壟斷
- **non-transferable** 不可轉讓的
- **upon the receipt of...** 收到…後
- **prior** 事前的
- **commence** 開始
- **in full force and effect** 具完整效力

- **automatically** 自動地
- **renew** 延長、更新
- **non-renewal** 不更新的
- **expiration** 到期、期滿
- **royalty** 版稅
- **undersigned** 簽名者、簽署人

美國百大企業文書撰寫祕訣 ！

Licensing contract（授權合約）是當事人允許另一方使用自己資產的證明文件。一般這樣的合約代表授權者（licensor）允許被授權者（licensee）使用自己特定的商標、圖像、資訊、技術、物品、想法等，而合約最核心的項目便是被授權者須支付給授權者的「報酬」。報酬通常都是以使用費、版稅（royalty）、租用費支付。

實務負責人的一句話 ！

授權合約在法律上有許多具爭議性的案例，所以製作十分複雜。尤其財政部分和授權期間必須具體明示，之後才能預防訴訟。

文書架構詳細解析

COPYRIGHT LICENSE AGREEMENT

❶ This LICENSE AGREEMENT (the "Agreement") is made and entered into effective as of May 15, 20×× (the "Effective Date"), by and between ART TECHNOLOGIES (the "Licensor") and MASON CO. (the "Licensee").

❷ Grant of License. In accordance with the terms and conditions of this Agreement, Licensor hereby grants to Licensee a non-exclusive, non-transferable license to use the Work in the course of its business and to otherwise copy, make, use and sell the Work, and for no other purpose. Any other use shall be made by Licensee only upon the receipt of prior written approval from Licensor.

❸ Term and Termination. This Agreement shall commence as of the Effective Date and shall continue in full force and effect for a period of two (2) years, and shall automatically renew for additional one year periods, unless either party provides written notice of non-renewal to the other party, not less than sixty (60) days prior to the expiration of any one (1) year term.

❹ Fees. Licensee shall pay the Licensor a royalty of five percent (5%) of gross receipts from sale of the Work each month. All royalties are to be paid within ten (10) days of the end of each month.

❺ IN WITNESS WHEREOF, the undersigned have executed this Agreement as of the date first above written.

Licensor
ART TECHNOLOGIES
_____

Licensee
MASON CO.
_____

❶ **標題和當事人（Title and Parties）**
以 Licensing Contract、License Agreement 等作為標題，並寫上合約的當事人。

❷ **定義和範圍（Definition and Scope）**
定義和本授權相關的著作物及著作權等，以及寫下授權的使用範圍。

❸ **期間和終止（Term and Termination）**
清楚寫出本授權合約的有效期間和合約終止條款。

❹ **費用（Fees）**
寫上 licensee 必須支付給 licensor 的授權費用，大部分都是以使用費、版稅（royalty）、租用費支付。

❺ **簽名（Signature）**
為保障合約的法律效力，合約當事人雙方都必須簽名。

撰寫文書所需的主要用語

01 **commence as of...** 自…（日期）生效
This agreement shall **commence as of** January 1, 20✕✕.
本合約自 20✕✕ 年 1 月 1 日起生效。

02 **continue in full force and effect** 繼續保有完整的效力
The provisions of this Agreement shall **continue in full force and effect** for one year.
本合約之條款一年內將持續保有完整的效力。

03 **automatically renew** 自動更新
The license will **automatically renew** unless either party says otherwise.
除非任一方有異議，否則授權將自動更新。

04 **provide written notice of non-renewal** 以書面通知不再續約
The licensor must **provide a written notice of non-renewal** for the license to be terminated.
授權方必須以書面通知將不再續約，才可終止授權。

05 **own all right, title and interest** 擁有所有權利、擁有權及利益
The Company **owns all right, title and interest** in the Work.
公司擁有著作物相關之所有權利、擁有權及利益。

06 **agree to be bound by the terms of...** 同意受…條款之約束
I hereby **agree to be bound by the terms of** this license.
本人同意受此授權條款之約束。

07 **be authorized to use...** 被授權使用…
You **are authorized to use** the Software for personal purposes.
基於個人使用目的，您可以使用該軟體。

08 **reserve the right to...** 保留…權利
Wizen **reserves the right to** terminate the license at any time.
威紳保留隨時終止授權的權利。

小測驗

01. The license will ( ) every two years.
    本授權每兩年將自動更新。

02. This agreement shall ( ) of July 1, 20✕✕.
    本合約自 20✕✕ 年 7 月 1 日起生效。

03. You ( ) to use the X-Men characters.
    您有權使用 X 戰警的角色。

04. The licensor ( ) to claim 30% of all profits.
    著作權人保留要求所有收益 30% 的權利。

160

END-USER LICENSE AGREEMENT

By installing, copying, accessing or otherwise using the ITSS software version 2.1 (the "ITSS SOFTWARE"), you (the "END USER") agree to be bound by the terms of this End-User License Agreement ("EULA"). If you do not agree to these terms, do not install, copy, access, or otherwise use the ITSS SOFTWARE.

1. AGREEMENT:

This EULA is a legal agreement between the END USER and Sidtech Inc. ("SIDTECH") who owns all right, title and interest in the ITSS SOFTWARE and related Documentation, files and intellectual property.

2. SOFTWARE:

The ITSS SOFTWARE includes any associated media, printed materials and electronic documentation ("Documentation") and any software updates, add-on components, web services and/or supplements that SIDTECH may provide to you or make available to you.

3. GRANT OF RIGHTS:

The ITSS SOFTWARE is licensed to you, not sold. You are authorized to use the ITSS SOFTWARE as follows: SIDTECH grants you a non-transferable and non-exclusive right to use a registered copy of the ITSS SOFTWARE for non-commercial purposes.

4. TERMINATION:

SIDTECH reserves the right to terminate your license and consequently your use of the ITSS SOFTWARE at any time. Your license will also automatically terminate if you fail to comply with any term or condition of this EULA. You may also terminate this license at any time. In the event you wish to terminate your license to use the ITSS SOFTWARE, you must inform SIDTECH of your intention. You agree, on termination of this license, to destroy all copies of the ITSS SOFTWARE and Documentation in your possession.

單字

- install 安裝
- access 讀取
- end-user 最終用戶
- legal agreement 法律合約、合法協定
- intellectual property 智慧財產
- electronic documentation 電子文書
- add-on component 附加成分
- grant of right 授權
- automatically 自動地
- comply with... 遵守…
- destroy 毀壞
- in your possession 你擁有的

保密合約

文書範本 1 保密合約 ▶翻譯請參閱 p.258

NON-DISCLOSURE AGREEMENT

This Non-Disclosure Agreement ("Agreement") is entered into and made effective as of the date set forth below, by and between Skynet Corporation ("Disclosing Party"), and Telpet Company ("Receiving Party").

1. **Confidential Information.** The confidential, proprietary and trade secret information being disclosed by the Disclosing Party is that information marked with a "confidential", "proprietary", or similar legend.

2. **Obligations of Receiving Party.** The Receiving Party will maintain the confidentiality of the Confidential Information of the Disclosing Party with the same degree of care that it uses to protect its own confidential and proprietary information. The Receiving Party will not disclose any of the Disclosing Party's Confidential Information to employees or to any third parties.

3. **Time Period.** The Disclosing Party will not assert any claims for breach of this Agreement or misappropriation of trade secrets against the Receiving Party arising out of the Receiving Party's disclosure of Disclosing Party's Confidential Information made more than three (3) years from the date of receipt of the Confidential Information by the Receiving Party.

This Agreement and each party's obligations shall be binding on the representatives, assigns, and successors of such party. Each party has signed this Agreement through its authorized representative.

Disclosing Party   Receiving Party
Signature _____   Signature _____
Printed Name _____   Printed name _____
Date _____   Date _____

單字

- **proprietary** 專利的、所有人的
- **legend** 範例
- **breach** 違反
- **binding on...** 有約束力的
- **assign** 指名人
- **successor** 繼承人

美國百大企業文書撰寫祕訣 ！

Non-disclosure agreement（保密合約 (NDA)）是作為保護資訊且具有法律約束力的文件，內容將闡述當事人雙方共享的資訊。本合約最大的目的在於保密，避免將資料洩漏給第三方，企業和職員、企業和企業之間皆可簽訂。若為企業之間的保密合約，便會針對雙方的合作投資或共同計畫等狀況，撰寫保護彼此的資訊、技術、商業機密為目的之合約。

實務負責人的一句話 ！

製作保密合約時，必須清楚明示合約當事人不可洩漏給第三方的機密資訊為何，且須一併寫上保密期間，以及違約時將採取的法律措施。

文書架構詳細解析

NON-DISCLOSURE AGREEMENT

❶ This Non-Disclosure Agreement ("Agreement") is entered into and made effective as of the date set forth below, by and between Skynet Corporation ("Disclosing Party"), and Telpet Company ("Receiving Party").

❷ 1. **Confidential Information.** The confidential, proprietary and trade secret information being disclosed by the Disclosing Party is that information marked with a "confidential", "proprietary", or similar legend.

❸ 2. **Obligations of Receiving Party.** The Receiving Party will maintain the confidentiality of the Confidential Information of the Disclosing Party with the same degree of care that it uses to protect its own confidential and proprietary information. The Receiving Party will not disclose any of the Disclosing Party's Confidential Information to employees or to any third parties.

❹ 3. **Time Period.** The Disclosing Party will not assert any claims for breach of this Agreement or misappropriation of trade secrets against the Receiving Party arising out of the Receiving Party's disclosure of Disclosing Party's Confidential Information made more than three (3) years from the date of receipt of the Confidential Information by the Receiving Party.

This Agreement and each party's obligations shall be binding on the representatives, assigns, and successors of such party. Each party has signed this Agreement through its authorized representative.

❺ Disclosing Party   Receiving Party
Signature _____   Signature _____
Printed Name _____   Printed name _____
Date _____   Date _____

❶ **標題和當事人（Title and Parties）**
以 Non-disclosure、Confidentiality Agreement 等作為標題，並寫上合約的當事人。

❷ **保密資訊（Confidential Information）**
清楚定義機密資訊為何，並一併規範機密資訊的範疇。

❸ **資訊收受方的義務（Obligations）**
具體敘述資訊收受方對機密資訊的內容須盡的義務，例如不得將合約所提及之機密資訊洩漏給職員或第三方等人。

❹ **期間（Time Period）**
記錄本合約提及之機密資訊的保密期間。

❺ **簽名（Signature）**
為保障合約的法律效力，合約當事人（代理人）雙方都必須簽名。

01 **maintain the confidentiality of...** 對…加以保密
The Receiving Party will **maintain the confidentiality of** the said information.
資訊收受方將對前述資訊加以保密。

02 **disclose confidential information** 洩漏機密資訊
Neither party shall **disclose confidential information**.
任一方皆不得洩漏機密資訊。

03 **shall be binding on...** 應具有…效力
The agreement **shall be binding on** the parties and their assigns.
本合約對當事人及其受讓人具有效力。

04 **protect the secrecy of...** 保護…的機密
The Receiving Party is to **protect the secrecy of** the documents at all costs.
資訊收受方無論如何皆應保護文件機密。

05 **be in compliance with...** 遵守…
You are to **be in compliance with** this Confidentiality Agreement.
您必須遵守本保密合約。

06 **recover liquidated damages** 追討違約金
PS Group is entitled to **recover liquidated damages**.
PS 集團有權追討違約金。

07 **refer to... by a code name** 以代號提及…
You are **to refer to** the Disclosing Party **by a code name**.
您必須以代號提及資訊揭露方。

08 **take appropriate legal actions** 採取適當的法律措施
The Disclosing Party will **take appropriate legal actions**.
資訊披露方將採取適當的法律措施。

小測驗

01. You are to refer to the Disclosing Party by a (   ).
    您必須以代號提及資訊揭露方。

02. Otherwise Comcast will (   ) legal actions.
    否則康卡斯特公司將採取適當的法律措施。

03. The Receiving Party is to (   ) of the following information.
    資訊收受方應保護如下資訊的機密。

04. This agreement (   ) on the parties and their assigns.
    本合約對當事人及其受讓人具有效力。

Restricted Project Agreement

You, the undersigned company, are involved in the development of a highly confidential Cantech project code named Emerald. This Cantech Restricted Project Agreement describes requirements that are intended to protect the secrecy of the project.

1. You are not to disclose any confidential information related to the project to any personnel other than those that have been expressly approved by Cantech to access such information.

2. You agree to require all approved-personnel to sign the confidentiality form prior to receiving confidential information about the project.

3. You agree to allow Cantech to verify that you are in compliance with this Agreement and other related agreements by auditing your records and information systems, inspecting your facilities, and interviewing your personnel.

4. Any violation of this Agreement will entitle Cantech to recover liquidated damages as set forth previously between Cantech and you, dated September 22, 20××.

Security Requirements:

1. Establish a security team with a qualified security manager.
2. Refer to Cantech by a code name.
3. Refer to the restricted project by its code name.
4. Conduct regular training regarding confidentiality and security.
5. Notify Cantech of any unauthorized exposures, thefts, or losses of confidential information and materials, and take appropriate legal actions and remedies.

單字

- intend 意圖
- secrecy 秘密
- personnel 員工、職員
- verify 證明、證實
- in compliance with... 根據…、遵循…
- audit 查帳、審計
- inspect 審查

- violation 違反
- recover 挽回、彌補（損失等）
- liquidated damage 損壞賠償、違約金
- exposure 揭發、揭露
- appropriate 適當的、洽當的
- legal action 法律措施
- remedy 療法、補救（辦法）

文書範本 1 不動產租賃契約 ▶翻譯請參閱 p.260

COMMERCIAL LEASE AGREEMENT

This LEASE AGREEMENT is made and entered into on January 23, 20××, by and between Candlestick, LLC, (hereinafter referred to as "Landlord"), and Teller Energy, Inc., (hereinafter referred to as "Tenant").

PREMISES. Landlord, in consideration of the lease payments provided in this Lease, leases to Tenant the fourth floor of Sungjil Building located at 78 World Cup-ro, Mapo-gu, Seoul. (the "Property")

LEASE TERM. The lease will begin on January 25, 20×× and will terminate on January 24, 20××. This Lease shall automatically renew for an additional period of two (2) years per renewal term, unless either party gives written notice of termination no later than 30 days prior to the end of the term or renewal term.

LEASE PAYMENTS. Tenant shall pay to Landlord monthly installments of 10,000,000 won, payable in advance on the fifth day of each month.

SECURITY DEPOSIT. At the time of the signing of this Lease, Tenant shall pay to Landlord, in trust, a deposit of 200,000,000 won to be reimbursed at the end of the Lease.

MAINTENANCE AND UTILITIES. Landlord shall have the responsibility to maintain the Property in good repair at all times. Tenant shall be responsible for all utilities incurred in connection with the Property.

Landlord:
Candlestick, LLC

Tenant:
Teller Energy, Inc.

_____

_____

單字

- tenant 承租人、租客
- premises （包括附屬建築、土地等在內的）不動產
- automatically renew 自動更新
- renewal term 更新期間
- monthly installment 月租
- payable in advance 可預先支付
- pay in trust 信託、委託
- reimburse 補償、賠償
- be in good repair 維修良好、完好無損
- utilities （水電瓦斯等）公共費用
- incur 招致、使…發生

美國百大企業文書撰寫祕訣！

Lease agreement（租賃契約）是出租人和承租人之間所締結的租賃合約。企業間主要簽的是辦公空、店面之類的商業租賃契約，稱做 commercial lease（商業租契）。租賃契約基本上包含當事人（出租人、承租人）的資訊、不動產地址及租賃期間、租金、保證金等資料。公共費用該由誰承擔也可以一併寫入合約中。除了不動產，也有影印機、飲水機等的設備租賃契約（equipment lease）。

實務負責人的一句話！

因為租賃合約為法律文件，撰寫時須以清楚明瞭為主。除了租賃不動產、設備資料、租賃條件和租金之外，也要明確寫下租客必須遵守的條款。

文書架構詳細解析

COMMERCIAL LEASE AGREEMENT

❶ This LEASE AGREEMENT is made and entered into on January 23, 20××, by and between Candlestick, LLC, (hereinafter referred to as "Landlord"), and Teller Energy, Inc., (hereinafter referred to as "Tenant").

❷ PREMISES. Landlord, in consideration of the lease payments provided in this Lease, leases to Tenant the fourth floor of Sungjil Building located at 78 World Cup-ro, Mapo-gu, Seoul. (the "Property")

❸ LEASE TERM. The lease will begin on January 25, 20×× and will terminate on January 24, 20××. This Lease shall automatically renew for an additional period of two (2) years per renewal term, unless either party gives written notice of termination no later than 30 days prior to the end of the term or renewal term.

LEASE PAYMENTS. Tenant shall pay to Landlord monthly installments of 10,000,000 won, payable in advance on the fifth day of each month.

❹ SECURITY DEPOSIT. At the time of the signing of this Lease, Tenant shall pay to Landlord, in trust; a deposit of 200,000,000 won to be reimbursed at the end of the Lease.

❺ MAINTENANCE AND UTILITIES. Landlord shall have the responsibility to maintain the Property in good repair at all times. Tenant shall be responsible for all utilities incurred in connection with the Property.

Landlord:	Tenant:
❻ Candlestick, LLC	Teller Energy, Inc.
_____	_____

❶ 標題和當事人（Title and Parties）
以 Lease Agreement 作為標題，並寫上出租人和承租人資訊。

❷ 資產（Property）
具體寫下租賃不動產或租賃設備的相關資訊。

❸ 租賃期間（Lease Term）
設定不動產或設備的租賃期間。

❹ 租金和保證金（Lease Payments and Security Deposit）
寫下租金及保證金的金額，同時一併寫明支付租金的時間點及支付方式。

❺ 維修和公共費用（Maintenance and Utilities）
若為不動產租賃契約，則應明示維修和公共費用由哪一方負擔。

❻ 簽名（Signature）
合約當事人皆需簽名。

撰寫文書所需的主要用語

01 **the lease will begin/start on...**　租賃期限將始於…（日期）
The lease will begin on January 1, 20××.
租賃期限將始於 20×× 年 1 月 1 日。

02 **give written notice of termination**　以書面通知終止契約
The tenant must **give written notice of termination** 60 days in advance.
承租人須於 60 日前以書面通知終止契約。

03 **prior to the end of the term**　租約到期前
The lessee must return the property **prior to the end of the term**.
承租人須於租約到期前歸還資產（不動產）。

04 **be responsible for...**　負責…
The tenant shall **be responsible for** all utilities.
承租人須負擔所有公共費用。

05 **lease the following equipment to...**　出租以下設備給…
The Lessor **leases the following equipment to** the Lessee.
出租人出租以下設備給承租人。

06 **pay monthly installments of...**　每月分期支付…
The lessee shall **pay** to the lessor **monthly installments of** $500.
承租人須每月分期支付 500 美金。

07 **take possession of...**　接管…
The lessee may **take possession of** the equipment after July 31.
承租人於 7 月 31 日後，即可接管該設備。

08 **at the expiration of the lease term**　租期屆滿時
The property shall be vacated **at the expiration of the lease term**.
租期屆滿時，須搬出租賃之房屋。

小測驗

01. The Lessee shall (　　　　　　　　　　) of the copy machine on March 10, 20××.
承租人於 20×× 年 3 月 10 日取得影印機。

02. The lease (　　　　　　　　　) January 20, 20××.
租賃期限將始於 20×× 年 1 月 20 日。

03. The tenant shall (　　　　　　　　　　) for all utilities.
承租人須負擔所有公共費用。

04. The building shall be vacated (　　　　　　　　　　　) of the lease term.
建築物須於租期屆滿時完全清空。

Equipment Lease Agreement

This Equipment Lease Agreement ("Agreement") is made and entered on April 15, 20××, by and between IntranetIt ("Lessor") and TechMode.com ("Lessee") (collectively referred to as the "Parties").

1. **EQUIPMENT:** Lessor hereby leases to Lessee the following equipment: Two (2) photocopiers (Model: Xerox WorkCentre 7970) ("Equipment")

2. **LEASE TERM:** The lease will start on April 20, 20×× and will end on April 19, 20××. ("Lease Term")

3. **LEASE PAYMENTS:** Lessee agrees to pay to Lessor as rent for the Equipment the amount of $250.00 ("Rent") each month in advance on the first day of each month at 3392 Shadowmar Drive, Metairie, Louisiana. If any amount under this Agreement is more than 10 days late, Lessee agrees to pay a late fee of $100.

4. **SECURITY DEPOSIT:** Prior to taking possession of the Equipment, Lessee shall deposit with Lessor, in trust, a security deposit of $8,000.00 as security for the performance by Lessee of the terms under this Agreement and for any damages caused by Lessee or Lessee's agents to the Equipment under the Lease Term.

5. **POSSESSION AND SURRENDER OF EQUIPMENT:** Lessee shall be entitled to possession of the Equipment on the first day of the Lease Term. At the expiration of the Lease Term, Lessee shall surrender the Equipment to Lessor by delivering the Equipment to Lessor or Lessor's agents in good condition and working order, as it was at the commencement of the Agreement.

LESSOR:
IntranetIt

_____

LESSEE:
TechMode.com

_____

單字

- equipment 設備
- lessor 出租人
- lessee 承租人
- lease term 承租期間、租期
- in advance 預先、事先
- late fee 滯納金
- take possession 接管、占有
- agent 代理人
- security deposit 押金
- be entitled to... 有資格⋯
- surrender 繳還、放棄
- in good condition 良好狀態
- in working order 正常啟動、正常運作
- commencement 開始

169

文書範本 1　合作備忘錄　　　　　　　　　　　▶翻譯請參閱 p.262

MEMORANDUM OF UNDERSTANDING

This Memorandum of Understanding ("MOU") is made on 7 March 20×× by and between the Government of Malaysia represented by the Ministry of Communications and Multimedia (the "Government") and PT Netso Malaysia ("Netso"). (together, the "Parties").

Purpose. The purpose of this MOU is to provide the framework for future cooperation regarding the development of Information and Communication Technology ("ICT") in Malaysia between the Government and Netso.

Licensing and Partnership. The Parties confirm their intent that by no later than 31 October 20××, following completion of due process, the Parties will enter into binding contracts that licenses Netso software to the Government for use across all ministries, departments and agencies of the Government. Netso is to continuously support ICT projects identified by the Government that will improve access to information technology and develop a software economy in Malaysia.

Term and Termination. This MOU will commence on the day this MOU is signed and will continue in force until the earlier of: (i) the date of execution of a licensing transaction according to this MOU; and (ii) 31 October 20××.

Non-exclusivity and Confidentiality. This MOU is non-exclusive but the Parties must keep confidential the terms and conditions of this MOU.

IN WITNESS WHEREOF, the undersigned, acting as representatives of their respective Party, have signed this MOU.

Government of Malaysia　　　　　　　　　　　PT Netso Malaysia

_____　　　　　　　_____

單字

- framework 結構、體系
- cooperation 合作
- Information and Communication Technology 資訊通訊科技
- intent 意思、意向
- binding 有約束力的
- license 許可
- ministry 部門
- software economy 軟體產業
- continue in force 維持效力
- licensing transaction 授權交易
- according to... 根據…
- confidential 機密的

美國百大企業文書撰寫祕訣 ！

Memorandum of Understanding（MOU）是企業之間欲締結關係或形成合作關係時，簽署的協議書，例如短期項目或長期合作就可以簽署備忘錄。MOU 不只用於私人企業，政府機關和學校也經常需要簽署備忘錄。MOU 和一般合約不同，並不具法律約束力，但 MOU 和其他合約一樣，任一方都有履行的義務，且必須簽名。MOU 的優勢在於不受該國法律的限制。

實務負責人的一句話 ！

MOU 也像其他合約一樣，可以包含合約當事者的義務、合約期限、終止條件等，但大部分不包含支付款項的內容，但是支付款項的條款也還是能包含在 MOU 中。

文書架構詳細解析

MEMORANDUM OF UNDERSTANDING

❶ This Memorandum of Understanding ("MOU") is made on 7 March 20×× by and between the Government of Malaysia represented by the Ministry of Communications and Multimedia (the "Government") and PT Netso Malaysia ("Netso"). (together, the "Parties").

❷ Purpose. The purpose of this MOU is to provide the framework for future cooperation regarding the development of Information and Communication Technology ("ICT") in Malaysia between the Government and Netso.

❸ Licensing and Partnership. The Parties confirm their intent that by no later than 31 October 20××, following completion of due process, the Parties will enter into binding contracts that licenses Netso software to the Government for use across all ministries, departments and agencies of the Government. Netso is to continuously support ICT projects identified by the Government that will improve access to information technology and develop a software economy in Malaysia.

❹ Term and Termination. This MOU will commence on the day this MOU is signed and will continue in force until the earlier of: (i) the date of execution of a licensing transaction according to this MOU; and (ii) 31 October 20××.

❺ Non-exclusivity and Confidentiality. This MOU is non-exclusive but the Parties must keep confidential the terms and conditions of this MOU.

❻ IN WITNESS WHEREOF, the undersigned, acting as representatives of their respective Party, have signed this MOU.

Government of Malaysia   PT Netso Malaysia

_____   _____

❶ 標題和當事人（Title and Parties）
以 Memorandum of Understanding 等作為標題，並寫上 MOU 當事人的資訊。

❷ 目的（Purpose）
寫上 MOU 的目的。MOU 的目的可以是締結事業關係或業務合作等。

❸ 角色（Roles）
寫上 MOU 當事人須履行的角色、義務或責任等。

❹ 期間和終止（Term and Termination）
寫上 MOU 的期間和終止條件。

❺ 其他（Other）
寫下基本條款後，可以再追加是否壟斷、保密、其他責任等相關條款。

❻ 簽名（Signature）
合約當事人（代理人）皆需簽名。

01 **provide the framework for...**　提供⋯框架
This MOU **provides the framework for** the project in question.
本備忘錄提供與討論中計畫相關的框架。

02 **confirm their intent that/of...**　確認他們⋯的意向
The Parties **confirm their intent of** forming a long-lasting partnership.
當事人確定欲建立長期合作關係。

03 **following completion of due process**　完成合法程序後
Following completion of due process, the partners will draw up a legally binding contract.
合夥人完成合法程序後,將制定具法律約束力之合約書。

04 **keep (something) confidential**　將⋯保密
The Parties agree to **keep the content of the MOU** confidential.
當事人皆同意對 MOU 內容保密。

05 **act as representatives of...**　代表⋯行事
The undersigned are to **act as representatives of** their respective Parties.
簽署人即代表各方締約當事人。

06 **agree to work together**　同意一起合作
The University and Cisco **agree to work together** to promote new student programs.
該大學和希思科同意共同合作,推動新的學生課程。

07 **shall include but (are) not limited to...**　包括但不限於⋯
Responsibilities of the Company **shall include but are not limited to** the terms below.
公司的職責包括但不限於以下內容。

08 **services to be rendered by...**　由⋯提供的服務
Services to be rendered by the Company include planning of the event.
公司提供的服務包含本活動的規畫。

小測驗

01. This MOU (   ) for the joint project.
本備忘錄提供與該共同項目相關的框架。

02. The parties (   ) together.
當事人都同意一起合作。

03. (   ) of due process, a legally binding contract will be drawn up.
完成合法程序後,將制定具法律約束力之合約書。

04. Services to be (   ) by Aetna include the following.
安泰納公司提供的服務包含如下。

Memorandum of Understanding

This Memorandum of Understanding (the "Memorandum") is made on March 10, 20××,

BETWEEN ADT Corporation ("ADT")
AND　　 FT Korea ("FT"),
AND　　 Boston Innovation Center ("Center"). (together, the "Partners")

Purpose and Obligations

The Partners hereby agree to jointly hold the 20×× Innovation Cup (the "Event") to be held on May 15-20, 20×× in Boston, Massachusetts. The Partners acknowledge that no contractual relationship is created between them by this Memorandum, but agree to work together in the true spirit of partnership to ensure a successful event.

Cooperation

The responsibilities and services for the Event shall include, but not limited to:

a. Services to be rendered by ADT include: Preparation of all activities pertaining to the Event including registration of participants.
b. Services to be rendered by FT include: Provision of all equipment and event related promotions.
c. Services to be rendered by Center include: Provision of venue and staff including volunteers.

Resources

ADT agrees to provide the financial and material resources in respect of the Event.

Term

The arrangements made by the Partners by this Memorandum shall remain in place from the signing of this Memorandum until the completion of the Event.

單字

- jointly 共同地
- hold 舉辦
- contractual relationship 合約關係
- work together 合作
- true spirit 真正的精神
- render 提供
- registration 登記
- participant 參加者

- promotion 宣傳
- venue 場所
- volunteer 志願者
- financial resources 經濟支援
- material resources 物資支援
- in respect of... 關於⋯的
- arrangement 準備、安排、約定
- remain in place 維持

文書範本 1 購入房產意向書

▶翻譯請參閱 p.264

LETTER OF INTENT

Dear Schema Company,

The following Letter of Intent ("LOI") outlines the general terms under which Caskell Inc. ("Purchaser") is willing to purchase the below Property from Schema Company ("Seller").

1. **Property.** The real property known as "High Tower" (the "Property") consisting of approximately 12,000 square feet of warehouse and office space, located at 1486 School Street, Wilton, Connecticut.

2. **Purchase Price.** The purchase price for the Property shall be $15,000,000 (the "Purchase Price") in cash at closing.

3. **Due Diligence.** Purchaser shall have thirty (30) days from the execution of this Letter of Intent to perform all of its due diligence with respect to the Property, including making a physical inspection of the Property, in order to enter into a legally binding definitive Purchase and Sale Agreement.

4. **Acceptance.** If the terms of this Letter of Intent are acceptable, please indicate your acceptance by signing below and returning one original Letter of Intent to the undersigned.

Sincerely yours,

PURCHASER:

_____

Agreed to and Accepted this _____ day of _____, 20_____:

SELLER:

_____

單字

- outline 概述
- general terms 一般條款
- real property 不動產
- square feet 平方英尺
- at closing 交易結束時
- execution 實行、履行
- due diligence 審慎評估
- physical inspection 實地檢查
- legally binding 具法律效力
- definitive 最終的
- acceptable 可接受的
- original 原本的

美國百大企業文書撰寫祕訣！

Letter of intent（LOI）是企業間在正式簽訂合約前，聲明各自立場和意願的文件，和備忘錄（MOU）一樣不具法律約束力。LOI 通常會在合併、收購、合作投資、不動產租賃等交易時簽訂，目的在於為簽署人謀取利益。意向書沒有固定的格式，但同合約一樣，會以條款的方式呈現，文件最上方則會以書信的格式起頭。

實務負責人的一句話！

當可能簽訂事業合約時，為了保障當事人的權利和意願，建議將雙方洽談的內容書面化。意向書為將事業意圖及結果寫成書面後，雙方共同持有的文件。

文書架構詳細解析

❶ **LETTER OF INTENT**

Dear Schema Company,

❷ The following Letter of Intent ("LOI") outlines the general terms under which Caskell Inc. ("Purchaser") is willing to purchase the below Property from Schema Company ("Seller").

1. Property. The real property known as "High Tower" (the "Property") consisting of approximately 12,000 square feet of warehouse and office space, located at 1486 School Street, Wilton, Connecticut.

2. Purchase Price. The purchase price for the Property shall be $15,000,000 (the "Purchase Price") in cash at closing.

❸ **3. Due Diligence.** Purchaser shall have thirty (30) days from the execution of this Letter of Intent to perform all of its due diligence with respect to the Property, including making a physical inspection of the Property, in order to enter into a legally binding definitive Purchase and Sale Agreement.

4. Acceptance. If the terms of this Letter of Intent are acceptable, please indicate your acceptance by signing below and returning one original Letter of Intent to the undersigned.

Sincerely yours,

PURCHASER:

_____

❹ Agreed to and Accepted this _____ day of _____, 20____:

SELLER:

_____

❶ **標題（Title）**
以 Letter of Intent 等作為標題。

❷ **當事人（Parties）**
寫下意向書的當事人。因為格式類似某一方當事人寄給另一方當事人的信，所以中間會包含「Dear...」的內容。

❸ **主要內容（Main Body）**
意向書的寄件人會寫下欲向收信人提議的內容，可以包含當事人的義務、合約的潛在條件等。例如和買賣合約相關的意向書，便可以一起寫下買賣交易的意願，以及和買賣物品相關的內容。

❹ **簽名（Signature）**
制定意向書的當事人先簽名，並留下收信人簽名的空間。

01 **outline the general terms**　概述一般條款
This LOI **outlines the general terms** of a possible formal agreement.
本意向書概述正式合約中的一般條款。

02 **perform due diligence**　審慎評估
The board will **perform due diligence** to assess the situation.
董事會將會審慎評估狀況。

03 **enter into a legally binding agreement**　簽訂具有法律約束力的合約
The Parties will **enter into a legally binding agreement** after due diligence.
經過審慎評估後，雙方將簽訂具有法律約束力的合約。

04 **if the terms are acceptable**　若能接受上述條款
If the terms of this LOI are acceptable, sign below and forward a copy.
若您能接受本意向書中的條款，請以於下方簽名後回傳副本。

05 **indicate your acceptance by...**　透過…表示接受
Indicate your acceptance of this document **by** signing below.
若您接受本文件，請於下方簽名。

06 **it is the intention of... to...**　做…為…的目的
It is the intention of this Letter **to** form a basis of agreement.
本意向書的目的為形成合約的基礎。

07 **award the contract to...**　和…簽約
We hope to **award the contract to** you by the end of the month.
我們希望這個月底能與您簽約。

08 **form an interim agreement**　構成臨時合約
This Letter of Intent will **form an interim agreement** between the Parties.
本意向書將構成當事人之間的臨時合約關係。

小測驗

01. We hope to (　　　　　　　　　) to you.
    我們希望能與您簽約。

02. If the terms (　　　　　　　　　) sign below.
    若您能接受上述條款，請於下方簽名。

03. The LOI (　　　　　　　　　) of a possible contract.
    本意向書概述合約中的一般條款。

04. Indicate (　　　　　　　　　) by signing below.
    若您接受，請於下方簽名證明。

LETTER OF INTENT

Dear Mr. Kendall,

It is the intention of Wirtzi Co. to award the contract for the construction of our new plant in South Chungcheong to Lofta Construction. In addition, we wish to have you start work prior to the finalization of contract documents and final project pricing.

Under no circumstances shall you undertake any work nor make any commitments such that the total value of the work performed or commitments made pursuant to the instructions contained in this letter is greater than $5,000,000. The scope of work to be undertaken pursuant to this letter is:

• Site layout and survey;
• Acquisition of necessary permits and insurances sufficient to enable the start of construction;
• Mobilization and installation of temporary construction offices

This letter will form an interim agreement between Wirtzi Co. and Lofta Construction until such time as the ongoing negotiations have concluded with either a formal CCDC-2 contract being executed or an agreement that the parties will discontinue working together on the project.

Yours truly,

Wirtzi Co.

_____

Agreed to all terms and conditions above on March 1, 20××.

Lofta Construction

_____

單字

- intention 意圖、意思、目的
- finalization 最終化、最終認可
- under no circumstances 無論在任何狀況下
- undertake 開始做…
- pursuant to... 依照…、按照…
- layout 配置、排列
- survey 調查
- permit 許可證
- mobilization 動員
- installation 安裝、設置
- interim 暫時、過渡期
- ongoing 正在進行的
- execute 執行、履行
- discontinue 中斷

合作協議

文書範本1 合作協議　　　　　　　　　　　　　▶ 翻譯請參閱 p.266

PARTNERSHIP AGREEMENT

THIS AGREEMENT is made and entered into at July 6, 20×× by and between Poexod Portal and Hub Airlines (hereafter collectively referred to as the "Partners).

Partnership Name and Purpose. The Partners hereby form a Partnership under the name of "Poexod-Hub Online Service" to provide online services for travelers to purchase Hub Airlines tickets.

Term. The Partnership shall begin on July 10, 20×× and will continue until terminated by either Party.

Contributions. Each Partner shall contribute the following:

Poexod Portal: Website related services and online promotion.

Hub Airlines: Travel service, travel schedule, ticket packages and customer service.

Capital Contributions. Each Partner shall contribute to the Partnership, an initial contribution of $10,000.00. Each Partner shall share in the net annual operating profits or losses in the ratio of 1:1.

Withdrawal and Dissolution. The Partners hereby reserve the right to withdraw from the Partnership at any time. The Partnership may be dissolved by majority vote. All funds after debts have been paid will be distributed based on the ratio of 1:1.

IN WITNESS WHEREOF, the Partners hereto set their hands and seals the date first above mentioned.

Signature _____　Date_____
Signature _____　Date_____

單字

- partnership 合夥、夥伴
- traveler 旅客
- contribution 貢獻
- initial 一開始、初期
- annual 一年的、每年的
- operating profit 經營利潤
- loss 損失
- ratio 比例
- withdrawal 退出、撤回
- dissolution 解散、解除
- majority vote 多數決
- debt 債務

美國百大企業文書撰寫祕訣！

當想和其他當事人合夥或合作事業時，就會簽訂 Partnership Agreement（合作協議），內容包含合作進行的方式、各當事人須履行的義務等。適用於開發新事業、和新合夥人擴張現有事業。與此目的相似，但格式不同的還有為了加盟所需要簽訂的 Franchise Agreement（加盟協議），大部分是根據加盟店經營權所有人的要求來制定。

實務負責人的一句話！

合作協議或加盟合約最重要的是金錢問題，因此務必仔細確認各當事人須支付的投資金、利潤分配、權利金等。

文書架構詳細解析

PARTNERSHIP AGREEMENT

❶ THIS AGREEMENT is made and entered into at July 6, 20×× by and between Poexod Portal and Hub Airlines (hereafter collectively referred to as the "Partners).

❷ **Partnership Name and Purpose.** The Partners hereby form a Partnership under the name of "Poexod-Hub Online Service" to provide online services for travelers to purchase Hub Airlines tickets.

Term. The Partnership shall begin on July 10, 20×× and will continue until terminated by either Party.

❸ **Contributions.** Each Partner shall contribute the following:

Poexod Portal: Website related services and online promotion.

Hub Airlines: Travel service, travel schedule, ticket packages and customer service.

❹ **Capital Contributions.** Each Partner shall contribute to the Partnership, an initial contribution of $10,000.00. Each Partner shall share in the net annual operating profits or losses in the ratio of 1:1.

❺ **Withdrawal and Dissolution.** The Partners hereby reserve the right to withdraw from the Partnership at any time. The Partnership may be dissolved by majority vote. All funds after debts have been paid will be distributed based on the ratio of 1:1.

❻ IN WITNESS WHEREOF, the Partners hereto set their hands and seals the date first above mentioned.

Signature _____ Date_____

Signature _____ Date_____

❶ **標題和當事人（Title and Parties）**
以 Partnership Agreement 作為標題，並寫上協議當事人。

❷ **合作內容和期限（Partnership and Term）**
具體寫上合作的名稱、目的等基本事項，同時明示雙方合作的期限。

❸ **貢獻（Contributions）**
寫上合作關係成立時，各當事人應該提供的服務。

❹ **資本出資（Capital Contributions）**
寫上各當事人應出資的金額。

❺ **撤銷和解約（Withdrawal and Dissolution）**
敘述撤銷協議和解約的條件。

❻ **簽名（Signature）**
為保障本協議的法律效力，各當事人皆需簽名。

01 **form a partnership under the name of...** 以…名義建立合夥關係
The Partners are to **form a partnership under the name of** S&S.
合夥人將以 S&S 的名義建立合夥關係。

02 **each partner shall contribute...** 各合夥人需出資…
Each partner shall contribute 5,000 euros.
各合夥人需出資 5,000 歐元。

03 **share in the operating profits** 享營業額分潤
All parties will **share in the operating profits**.
所有當事人皆享營業額分潤。

04 **reserve the right to withdraw** 保留退出的權利
The Partners **reserve the right to withdraw** from the Partnership.
合夥人保留退出合夥的權利。

05 **be dissolved by majority vote** 以多數決終止
The contract can **be dissolved by majority vote**.
本合約可以多數決終止。

06 **pay an amount of... as franchise fee** 支付…作為加盟費
The Franchisee is to **pay an amount of** $5,000 **as a franchise fee**.
加盟經營者須支付 5,000 美元作為加盟費。

07 **pay a monthly royalty** 每月支付權利金
You are to **pay a monthly royalty** of $250 to the Franchisor.
您每月須支付特許人 250 美金的權利金。

08 **terminate upon the occurrence of...** 在發生…終止
The agreement can be **terminated upon the occurrence of** embezzlement.
本合約可於發生挪用公款時終止。

小測驗

01. Each partner (   ) to withdraw from the partnership.
各合夥人保留退出合夥的權利。

02. All parties will share in the (   ).
所有當事人皆享營業額分潤。

03. You are to (   ) of 500,000 won.
您每月須支付 500,000 韓元的權利金。

04. Each partner (   ) 25,000 dollars.
各合夥人需出資 25,000 美金。

Franchise Agreement

This Franchise Agreement is made on August 19, 20×× by and between Levine Coffee (the "Franchisor") and Roy Chung (the "Franchisee") on the basis of the following understandings and agreements:

1. **Grant of Franchise.** The Franchisor grants to the Franchisee, the right to the Franchisor's trademarks and proprietary methods of doing business in connection with the establishment and operation of a Levine Coffee franchise at 56 Choji-ro, Danwon-gu Ansan-si, Gyeonggi-do.

2. **Franchise Fee.** The Franchisee agrees to pay the Franchisor an amount of 20,000,000 won as Franchise Fee.

3. **Royalties.** Throughout the term of this Agreement, the Franchisee agrees to pay to the Franchisor a continuing monthly royalty equal to 8% of its Gross Sales generated from its franchise.

4. **Term.** The term of this Agreement begins on the date this Agreement is fully executed and ends two (2) years later. Either party may terminate upon thirty (30) days notice.

5. **Default and Termination.** The Franchisor shall have the right to terminate this Agreement upon the occurrence of any of the following events. a. Abandonment; b. Insolvency; c. Criminal Conviction, d. Failure to Make Payments; e. Misuse of Trademarks; f. Unauthorized Disclosure; g. Repeated Non-compliance.

This Agreement shall be signed on behalf of Franchisor by Darrel Skyman and on behalf of Franchisee by Roy Chung.

FRANCHISOR:
Darrel Skyman

FRANCHISEE:
Roy Chung

單字

- franchise 連鎖、加盟
- franchisor 特許人、加盟總部
- franchisee 被特許人、加盟主
- grant 授予
- trademark 商標權
- franchise fee 加盟金
- gross sales 總銷售額
- generate 招致、遭受
- default 契約未履行
- abandonment 放棄、拋棄
- insolvency 破產
- misuse 濫用
- unauthorized 未經授權的、未經許可的
- repeated 反覆
- non-compliance 違約行為

PART 6
Various Templates

各式文件範本

文書範本 1　估價單 Estimate　　　　▶翻譯請參閱 p.268

ESTIMATE

Date: September 24, 20××
Estimate: 48131311
Valid Until: September 30, 20××
Customer ID: PI719

TIEMA

To :　Pineup Corporation
　　　2303 Liberty Avenue
　　　Irvine, CA 92618
　　　714-422-0076

Quantity	Description	Unit Price	Line Total
20	Tiema A4 80 gsm Office Paper (500 sheets)	$9.50	$190.00
2	Basics Stapler (with 1,000 staples)	$4.50	$9.00

Subtotal	$	199.00
VAT Rate	%	7.50
VAT	$	14.90
Total	$	213.90

The above information is not an invoice and only an estimate of services/goods described above.
Payment will be collected in prior to provision of services/goods described in this quote.

To accept this quotation, sign here and return : _____

Thank you for your business!
Should you have any inquiries concerning this quote, please contact
Ronald A. Hensley (530-773-9081)
[Tiema Co., Ltd.] [2484 Byers Lane Redding, CA 96001]
Phone [530-773-9069] Fax [071-114-4932] [info@tiemapaper.com] [www.tiema.com]

單字

- valid until...　期限到…為止（效期）
- unit price　單價
- line total　該列總額
- VAT (Value Added Tax)　增值稅
- Invoice　發票
- collect　收（租金、稅金等）
- accept　接受
- quotation, quote　估價（單）
- inquiry　詢問
- concerning...　關於…

TIEMA

Invoice

Date: October 2, 20××
Invoice: INV-2013100207
Customer ID: PI719
Purchase Order: 4819
Payment Due By: October 24, 20××

Bill To:

Pineup Corporation
2303 Liberty Avenue
Irvine, CA 92618
714-422-0076

Ship To (If Different):

Pineup Corporation
2306 Liberty Avenue
Irvine, CA 92618
714-422-0081

Salesperson: Ronald A. Hensley   Delivery Date: October 4, 20××

Item	Description	Qty	Unit Price	Line Total
38601	Tiema A4 80 gsm Office Paper (500 sheets)	20	$9.50	$190.00
11381	Basics Stapler (with 1,000 staples)	2	$4.50	$9.00

Subtotal	$	199.00
VAT Rate	%	7.50
VAT	$	14.90
S&H	$	5.00
Discount	$	-2.50
Total	$	216.40

Please make all checks payable to Tiema Co.

Thank you for your business!
Should you have any inquiries concerning this invoice, please contact
Ronald A. Hensley (530-773-9081)
[Tiema Co., Ltd.] [2484 Byers Lane Redding, CA 96001]
Phone [530-773-9069] Fax [071-114-4932] [info@tiemapaper.com] [www.tiema.com]

單字

- purchase order 採購單
- salesperson 業務負責人
- delivery date 交貨日期
- Qty (Quantity) 數量

- S&H (Shipping and Handling) 運費及處理費
- discount 折扣
- check 支票
- payable to... 應支付給…

185

Smiant Ltd.

TAX INVOICE

3774 Terry Lane, Winter Park, FL 32789
Tel: (321) 303-5484, Fax: (321) 303-5480
Email: tax@smiant.com, Website: www.smiant.com

Issue Date: 10/20/20××

Invoice: 200003
P/O.: 05-BC-39138393

BILL To :
Prado Health
4746 Beech Street
Concord, CA 94520

Product ID	Description	Qty	UM	Unit Price	Amount
P48602	Wild blueberry	50	boxes	$25.00	$1,250.00

Subtotal	$ 1,250.00
Tax (10% rate)	$ 125.00
Invoice Total	$ 1,375.00
Freight	$ -
Amount Paid	$ -
Balance Due	$ 1,375.00

PAYMENT INFORMATION

Direct Deposit to:
Bank Name: EastWest
Account Name: Smiant Ltd.
Account No: 302597381981
Due Date: Within 20 days of issue of this invoice

單字

- issue 發行
- UM (Unit of Measure) 計量單位
- freight 運費
- amount paid 支付金額
- balance due 未付金額、結欠餘額
- payment information 付款資訊
- direct deposit 直接存款
- due date 到期日、（結帳）期限

SALES RECEIPT

Smiant Ltd.

3774 Terry Lane, Winter Park, FL 32789
Tel: (321) 303-5484, Fax: (321) 303-5480
Email: tax@smiant.com, Website: www.smiant.com

Date: 10/30/20××
Receipt: 98171
P/O.: 05-BC-39138393

SOLD TO:
Prado Health
4746 Beech Street
Concord, CA 94520

Product ID	Description	Qty	UM	Unit Price	Amount
P48602	Wild blueberry	50	boxes	$25.00	$1,250.00

Subtotal	$ 1,250.00
Tax (10% rate)	$ 125.00
Total	$ 1,375.00

Amount Received: $1,375.00
Payment Method: Cash
Sale Made By: Thomas Valle

Thank you for your business!

單字

- sales receipt 銷售發票
- amount received 收取金額
- payment method 結帳方式
- sale made by 銷售負責人

文書範本 1 採購訂單 Purchase Order ▶翻譯請參閱 **p.272**

Wirenet Computing
2338 Bee Street
Muskegon, MI, USA
Phone: (515) 910-7311 Fax: (515) 910-7350
Email: saunders@wirenet.com Website: www.wirenet.com

Wirenet
Computing

Purchase Order

No. : PO-03-187311

Date: March 3, 20✕✕

To : Raske Utilities
Weihburggasse 79
AICHBERG
Australia

Shipping Date	Shipping Terms	Payment Terms	Currency Code
March 13, 2017	CIF	60 Days	USD

Item No.	Description	Quantity	Unit	Unit Price	Amount
1413	Mouse Rubber Cap (BLUE)	5,500	PCS	4.50	24,750.00
	(VAT Inclusive)			**TOTAL**	**24,750.00**

Note.

1. Please send two copies of your invoice.
2. Enter this order in accordance with the price, terms, delivery method and specifications listed above.
3. Please notify us immediately if you are unable to ship as specified.
4. Late delivery will be subject to cancellation.

Prepared by
Shannon Saunders

Approved by
Russell C. Harper

單字

- terms 條款
- CIF (Cost, Insurance and Freight)
  成本、保險費加運費
- currency 貨幣
- inclusive 包括一切的

- copy （文書等的）副本
- enter an order 打訂單
- notify 通知
- unable 不能做、無法做
- be subject to... 可能因…而變動

COMMERCIAL INVOICE

Date of Exportation Mar-13-20××	Reference Invoice No. I-380033 AWB/BL No. 710-8371-9011
Terms of Sale CIF	
Shipper/Exporter Raske Utilities Weihburggasse 79 4707 AICHBERG Australia	Consignee Wirenet Computing 2338 Bee Street Muskegon, MI 48001 USA
Country of Origin of Goods Australia	Importer (If different than consignee)
Country of Ultimate Destination USA	

Full Description of Goods	Country of Manufacture	QTY (Units)	Unit Value	Total Value
Mouse Rubber Cap (BLUE) – Product No. 0901	Australia	5,500	USD 4.50	USD 24,750.00
Subtotal		5,500		USD 24,750.00

Total Number of Packages 55	Freight	0.00
	Insurance	0.00

These commodities, technology or software were exported from Australia in accordance with the Export Administration Regulations. Diversion contrary to Australian Law is prohibited.	Total Invoice Value USD 24,750.00

I hereby certify that the information on this invoice is true and correct and that the contents of this shipment are as stated above.

Signature of Shipper/Exporter
Edgar B. Lombardo

Date
March 13, 20××

單字

- exportation 出口
- reference 參照號碼
- AWB (Air Waybill) 航空貨運單
- BL (Bill of Lading) 託運單
- consignee 收貨人、收件人
- country of origin 原產地
- ultimate 最終

- country of manufacture 製造國
- freight 運費
- insurance 保險
- commodity 商品
- in accordance with... 根據…
- diversion 改變用途
- contrary to... 和…相反

PSK Line

Bill of Lading

Shipper	Packing List No.	Bill of Lading No.
Berry Berry 4765 West Side Avenue Newark, NJ 07102 USA TEL +1-229-371-3091	219476303-1	SSOF0376181
	Freight and Charges Payable By Shipper	
	Terms of Sale FOB	

Consignee	Number of Original B/L Issued
Apeech Inernational 49 Bullwood Rd Sproatley, HU11 3NZ UK TEL +44-814-391-9012	Three (3)
	Place and Date of Issue Newark, USA / November 12, 20✕✕

Notify Party	For Release of Shipment, Please Contact
Same as consignee	Plasme Agency

Place of Delivery	Place of Loading	No. of Containers	No. of Packages Received by the Carrier
Newark, USA	Newark, USA	1	200

Place of Receipt	Port of Discharge	For Transshipment to	Vessel/Voyage
Portsmouth, UK	Portsmouth, UK	-	LSPL UBRE / 0038E

Marks and Numbers	No. of PKGS	Description of Packages and Goods	Gross Weight	Measurement
20' Steel Dry Container No. POSL3084384	200 PKGS	200 PKGS of Wild Blueberry	900 KG	11.1 M³

Freight Prepaid
Shipped on Board: November 14, 20✕✕
Carrier Signature: *Lonnie J. Akers*

單字

- packing list 裝箱單
- freight and charges 運費
- payable by... …必須支付的
- FOB (Free on Board) 離岸價格、船上交貨價格
- release of shipment 卸貨
- receipt 領取
- place of delivery 交貨地、交付地
- transshipment 轉船
- vessel 船舶
- marks and numbers 運輸標誌
- PKGS (packages) 包裹
- gross weight 總重量

PACKING LIST

Shipped From	Shipped To
Cenntral Materials 3776 Taylor Street New York, NY 10007 USA	Indronestry Co. 1040 Station Road Pinetown 3624 South Africa Attention: Alan Monson

DATE ORDERED 11OCT12	ORDER NUMBER 967173	DATE OF SHIPMENT 21OCT12
SHIPPED VIA President Lines	CONTAINER NUMBER TA90198	INVOICE NUMBER BT-1923
TOTAL NUMBER OF PACKAGES 375	TOTAL GROSS WEIGHT 23,437.5 KGS	AWB/BL NUMBER TXV7483503120

No.	Item Number	Item Description	Quantity	Packaging Type	Per Package Gross Weight
1	Q18602	Polyester	125	Crate	87.5 KGS
2	Q18605	Polypropylene	250	Crate	50.0 KGS
3					
4					

Note: These commodities, technology or software were exported from the United States in accordance with the Export Administration Regulations.

Signature: *Connie D. Aguilar*   Date: *Oct. 21, 20*××
Connie D. Aguilar, Export Manager

單字

- via... 藉由⋯、透過⋯
- packaging type 包裝類型
- crate 箱子
- administration regulation 管理規定

文書範本 1 信用狀 Letter of Credit ▶ 翻譯請參閱 p.276

Statewide Bank

Issue Date: October 15, 20××
Irrevocable Standby Letter of Credit Number: 9113814

ISSUING BANK

Statewide Bank
3528 Hoffman Avenue
New York, NY 10016

BENEFICIARY	APPLICANT
Derivet Co. 204 Willison Street Minneapolis, MN 55415	Larrynx Group, Inc. 3124 Stoney Lonesome Road Bloomsburg, PA 17815

AMOUNT: Five hundred seventy three thousand nine hundred and 20/100 US dollars. (USD 573,900.20)
EXPIRATION: April 14, 20××

Ladies and Gentlemen:

At the request of Larrynx Group, Inc. ("Applicant"), we hereby establish our irrevocable standby letter of credit in your favor in the amount of five hundred seventy three thousand nine hundred and 20/100 US dollars (US$573,900.20). This letter of credit is available with us at our above office by payment against presentation of your draft(s) drawn on us at sight accompanied by the following.

1. Beneficiary's signed statement stating as follows:
   Derivet Co. certifies that Larrynx Group, Inc. has failed to complete the requirements for Project #AB391 and we are therefore entitled to the sum of US$573,900.20.
2. Copy of beneficiary's signed letter addressed to the applicant.

We hereby engage with you that the draft(s) drawn under, and in compliance with the terms of the credit, will be duly honored on presentation to us, on or before the expiration date.

Statewide Bank
James W. Cruz
Authorized Signature

James W. Cruz, Credit Manager
Name and Title of Authorized Signature

單字

- letter of credit 信用狀
- irrevocable 不可取消
- beneficiary 受益者，受惠者
- applicant 申請人，委託人
- expiration 到期，期滿

- draft （銀行發行的）匯票
- draw at sight 開立見票、即付匯票
- be entitled to... 有（接受）…的資格
- in compliance with... 根據…
- honor 履行、支付

文書範本 2 **資產明細表 Balance Sheet**

▶ 翻譯請參閱 p.277

Maise-Mart Stores Inc.
Balance Sheet
For Fiscal Year Ending December 31, 20××

Assets	
Current Assets	
Cash and Cash Equivalents	81,900,000
Short Term Investments	-
Inventory	59,833,000
Other Current Assets	14,398,000
Total Current Assets	**156,131,000**
Long Term Investments	201,390,000
Property Plant and Equipment	593,090,000
Intangible Assets	-
Other Assets	76,010,000
Total Assets	**1,026,621,000**
Liabilities	
Current Liabilities	
Accounts Payable	65,550,000
Short/Current Long Term Debt	15,330,000
Other Current Liabilities	1,300,000
Total Current Liabilities	**82,180,000**
Long Term Debt	36,700,000
Minority Interest	2,900,000
Other Liabilities	-
Total Liabilities	**121,780,000**
Stockholders' Equity	
Preferred Stock	-
Common Stock	16,900,000
Retained Earnings	46,990,000
Treasury Stock	5,550,000
Capital Surplus	3,500,000
Other Stockholders' Equity	-
Total Stockholders' Equity	**72,940,000**
Total Liabilities and Stockholders' Equity	**194,720,000**

單字

- balance sheet 資產明細表
- asset 資產
- short term 短期
- long term 長期
- current 流動
- liability 負債
- stockholders' equity 股東權益
- stock 股市、股市資本
- earning 所得，盈餘
- surplus 資金盈餘

Maise-Mart Stores Inc.
Income Statement
For Fiscal Year Ending December 31, 20××

Profit	
Net Sales	306,102,000
Other Income	1,900,000
Total Revenue	308,002,000
Cost of Goods Sold	248,654,000
Gross Profit	**59,348,000**

Expenses	
Research Development	-
Selling General and Administrative	21,180,000
Non-recurring	-
Others	5,330,000
Total Expenses	**26,510,000**
Operating Income (or Loss)	**32,838,000**

Earnings Before Taxes	32,838,000
Income Tax	4,712,000
Earnings Before Minority Interest	28,126,000
Minority Interest	230,000
Net Income	**27,896,000**

單字

- income statement 損益表
- net sales 銷售淨額
- gross profit 毛利、總收益
- research development 研究開發
- non-recurring 非經常性的

- operating income 營業利益
- operating loss 營業損失
- income tax 所得稅、法人稅
- minority interest 少數股東權益
- net income 淨利

Novascot

Date: Oct. 5, 20××

PAYSLIP

Employee Name: Seth Crisp
Designation: Financial Advisor

Employee No.: INT1091
Pay Period: September 20××

INCOME		DEDUCTIONS	
Basic Pay	1,600.00	Income Tax	101.45
Overtime Pay	-	Employee Pension	45.00
Bonus	400.50	Health Insurance	23.00
Medical Allowance	100.75		
Transport Allowance	200.00		
Food Allowance	110.00		
Other Allowance	-		
TOTAL INCOME	**2,411.30**	**TOTAL DEDUCTIONS**	**169.50**

NET PAY		2,241.80

Total Year to Date	
Taxable Gross Pay	20,801.70
Income Tax	896.40
Employee Pension	414.00

Paid by: Electronic Transfer
Bank: Citibank
Account No.: 0-163812-685

單字

- payslip, paystub 薪資明細
- designation 職銜
- income 收入
- basic pay 底薪
- allowance 津貼

- deduction 扣除額
- overtime pay 加班費
- pension 退休金
- taxable 需要納稅的
- electronic transfer 電子轉帳

THE USA

PART 7

APPENDIX.

翻譯及解答　**Translation & Answer**

文書範本 1

網頁服務開發工程師

第一軟體

德克薩斯州 奧斯汀

第一軟體正在徵求負責前端及使用者體驗開發的資深網路工程師。

負責工作內容

- 與其他 UX/UI 設計工程師密切合作，使用 HTML、CSS 和 JavaScript 等核心技術，進行產品開發、網路工程開發等，以利研發出完善的使用者應用程式。
- 持續升級前端系統，實現迅速的 UI 開發及疊代。
- 將業務程序及工作流程整理為共享文件，以便往後應用及再利用。

必要條件

- 具備 2 年以上前端、後端程式碼的審查及撰寫經驗。
- 精通 HTML、CSS 和 JavaScript。
- 至少熟悉 Node.js、Ruby on Rails 或 PHP 等後端程式碼之一。
- 熟悉 MongoDB 或 MySQL 等資料庫管理系統。
- 具備與專用託管服務業者合作，建構伺服器環境、維護及支援的經驗。

加分條件

- 在網頁設計上具備行動設備優先的思維。
- 具備用瀏覽器測試個人成品的習慣。
- 對產品使用者有同理心，並能預測使用者的需求。
- 熟悉 Photoshop、Illustrator 及 Sketch 等設計工具。

保險精算分析師

卡洛威保險公司

紐約州 紐約

工作性質：正職

經驗：入門等級

職務概述

卡洛威保險公司正在為紐約辦公室尋找入門等級的保險精算分析師。這是一個能夠廣泛接觸，各領域一流企業保險精算分析業務的特別機會。

工作內容

· 執行個人帳戶的定價分析。

· 分析數據及計算意外、財產損失、受傷、死亡等事件相關成本。

· 處理基層數據，編寫成報告。

· 支援整理精算數據及分析資料，以加強流程並維持高水準的數據完整性。

必要條件

· 具備大學學歷，或有精算科學、數學或統計學相關的同等培訓經歷

· 0 到 2 年的相關經歷

· 通過 1 項以上的保險精算考試

· 中級至高級 Excel 應用能力

· 優先考慮熟悉 SQL 者

· 出色的問題解決能力

· 紮實的口語及書面溝通能力

公司簡介

卡洛威於 1921 年起開始從事保險投資領域，現在在全球雇有 11,000 名以上的工作夥伴。

小測驗正解

01. is seeking　02. Familiarity　03. be exposed　04. unique opportunity

履歷

文書範本 1

馬克·C·克拉克

39501 佛羅里達州格爾夫波特康納街 2220 號 | 手機：(494) 7834-5684 | markclark@msn.com

---

網路內容管理經理

擁有 15 年線上行銷經驗的資深內容管理經理。為多家公司管理 150 個以上的網站。

· 訂定成長策略　　　　　　　　　· 傑出的合作能力
· 具備多媒體經驗　　　　　　　　· 專業跨媒體能力
· 主導電子商務　　　　　　　　　· 以服務客戶為導向

經歷

網路寶石公司　　　　　　　　　　**佛羅里達州傑克遜維爾市**
資深網路內容管理經理　　　　　　　**20×× - 現在**

· 前六個月網路流量增加 40%。
· 線上行銷成本降低 12%。
· 開發及設計公司內部網路。

網路顧問　　　　　　　　　　　　　**20×× - 20××**

· 為 22 間財星 500 大企業和多間新創公司，提供流量諮詢及改善服務。
· 與行銷團隊合作執行搜尋引擎最佳化（SEO）。

數位 A 公司　　　　　　　　　　　**佛羅里達州坦帕市**
網路分析師　　　　　　　　　　　　**20××- 20××**

· 分析線上行銷數據，以尋求新的商業機會。
· 執行 200 個以上的專業網站最佳化。

學歷

資訊科技，20××，科羅拉多大學，科羅拉多州波德市

技能

熟悉 CMS、Sitefinity、Ektron、XML 及 CSS

丹尼爾·阿爾梅達

daniel@maruchs.edu | 手機：(755) 7834-5684 | LINKEDIN.com/in/daniel-almeida

簡介

積極進取且以成果為導向，具備堅實的學歷和豐富的實習經驗，同時是充滿熱情且擁有傑出溝通能力的獨立專業人士。

核心特質

- 注重細節
- 以人為本
- 多工處理能力
- 優秀的溝通者

經歷

若谷廣告 **伊利諾州芝加哥市**
廣告業務實習生 **夏天 20××**

- 與業務團隊密切合作，制定策略。
- 協調廣播及平面媒體廣告活動。

MGT 中心 **伊利諾州春田市**
業務實習生 **夏天 20××**

- 在所有業務階段，與客戶交流。
- 致電潛在客戶（每天 50 通）。
- 為 50 個以上的客戶處理訂單及出貨。

學歷

馬魯奇學院 **伊利諾州芝加哥市** **20××-20××**
行銷 第二名（Magna Cum Laude）

活動

- 副會長，學生會，20××
- 總財務，貝塔伽馬榮譽會，20××-20-××

小測驗正解

01. Responsible for   02. Reduced costs   03. Proficient in   04. Built rapport

文書範本 1

葛雷哥萊·詹金斯

20741 馬里蘭州大學園查塔姆路 3326 號
手機：(522) 3737-4576
mjenkins@mail.com

肯·希爾曼先生鈞鑒，

我正在積極發展電腦網路領域的職涯，日前在安潔利斯特上注意到貴公司的網路工程師招聘公告，因此有意應徵這份職缺。

我認為吉歐戴瑞特的網路精簡模型，做到了跨時代的創新而且非常有趣，我對貴公司活用於網路基礎架構的最新硬體、安全技術有強烈的興趣。

我目前在華盛頓特區擔任網路工程師，具備排除故障並解決網路問題的經歷。解決問題是我的專長領域之一，希望未來能夠繼續從事該領域的工作。我相信我的專業技術，將會成為公司的重要資產。

隨信附上個人履歷以供貴公司審閱。若有任何疑問，歡迎與我聯絡。期待收到您的回信。

敬祝事事順利

葛雷哥萊·詹金斯敬上

金政煥

03930 首爾市麻浦區上岩路 397 號 2 樓
手機：010-3737-4576
junghwan173@mail.co.kr

人事主管鈞鑒，

我的名字是金政煥，很高興有機會應徵《金融時報》的初階記者職位。

我近期畢業於江南大學，在校內曾擔任校刊總編，而廣告業務是我眾多的工作之一，也是啟發我關注利潤和收益的契機。我學會開拓、管理及維繫與校園周邊商家間的業務關係。

我一直都很喜歡閱讀《金融時報》的建議專欄。其中收錄了許多開創性的建議，令我銘記在心。我已構思了許多個人獨有的建議，希望將來有機會為該專欄做出貢獻。

請務必參考我的履歷，以了解我的個人活動及成果。謝謝您撥冗閱讀此信。

敬祝萬事如意

金政煥 敬上

小測驗正解

01. currently working   02. great asset   03. am excited/thrilled
04. learned how to

推薦信

文書範本 1

李先生鈞鑒，

很榮幸向您推薦愛麗絲·柯蒂斯擔任貴公司的助理人事經理一職。愛麗絲·柯蒂斯曾與我在泰拉公司共事，我是她在 20×× 到 20×× 之間的經理及直屬主管。

由於她可靠且認真的工作態度，我和她一起在人力資源部門工作的經歷很愉快。她具備出色的人際處理能力，也有傑出的識人眼光。因為她的教育背景為我們提供強而有力的優勢，因此她負責員工訓練。她制定了一個有效的內部培訓計畫，至今我們仍用於提升員工團隊士氣。她是個真正具備團隊精神且能夠激勵周遭眾人的人物。

我非常推薦柯蒂斯小姐加入貴團隊，我相信她能帶給貴公司寶貴的助力。如對於她的能力和經歷有任何疑問，歡迎撥打 555-153-4557 或透過電子郵件與我聯絡。

敬祝事事順心

卡爾文·庫爾曼
人力資源部長
泰拉公司

三一大學，商管學院
申請入學組別—高階主管碩士在職專班（EMBA）
91766 加利福尼亞州波莫納市格倫代爾大道 4495 號

各位入學審查委員道鑒，

此次來信是為了向三一大學高階主管碩士在職專班，推薦德瑞士·蔡。我和德瑞士·蔡已相識六年，我們曾作為同事在位於首爾的 UP 太平洋分公司一起工作。

他是一名獨一無二的職員，因為他總是致力於建立與客戶之間的信任。因此在他於本店工作的過去三年間，本店在他的監督管理下創造了 110% 的銷售成長。在此之前，他也幫助本公司組織會計部門，為公司節省了 22,000 美元以上的年營業成本。

在我作為同事認識他的這些年來，他一向充滿活力，並抱持著樂觀及渴望學習的態度。我相信這些個人特質，能使他在貴校的在職專班成為出色的學生。如對本推薦信有任何疑問，歡迎隨時與我聯絡。

敬祝事事順心

黃瑞英
業務規劃經理
UP 太平洋分公司（韓國首爾）

小測驗正解

01. goes the extra mile　02. working with　03. to recommend　04. has an eye

文書範本 1

聖保羅健康中心

老年醫學顧問團隊   傳真封面
韋斯特蘭辦公室
明尼蘇達州聖保羅 9524 號
電話：070-393-4800
傳真：070-393-4880

www.stpaulhealthservices.org

日期：20×× 年 6 月 6 日   頁數：5
                                （含封面）

收件人	發件人
姓名：詹姆斯·E·沃爾登 傳真：070-7836-4880 電話：	姓名：邁克爾·洛扎 傳真：070-393-4880 電話：070-393-4844

訊息：（機密或個資請勿記於封面。）

致詹姆斯
傳真附上已簽名的合約影本。此傳真證明本公司同意您昨天提出的條件。如還有其他需要，請與我聯絡。

邁克爾

中網互助

88101 德克薩斯州克洛維斯艾瑪街 4210 號

電話：(350) 388-6931

傳真：(350) 388-6930

www.centralnetmutual.com

20×× 年 10 月 13 日

凱瑟琳・朗

哨兵新聞

32805 佛羅里達州奧蘭多奧卡拉街 3757 號

朗小姐台鑒，

函復您於 20×× 年 10 月 10 日就中網互助是否持有特定文件提出的詢問。

1. 20×× 年、20×× 年及 20×× 年度電子廢棄物處理及回收合約的書面影本。
   - **中網互助已沒有留存上述年份的合約影本。**
2. 20×× 年、20×× 年及 20×× 年度報廢硬碟處理合約的書面影本。
   - **中網互助已沒有留存上述年份的合約影本。**

如有其他需要協助的事項，請不吝告知。

敬祝事事順心

華萊士・阿莫斯

華萊士・阿莫斯

審計處處長

中網互助

小測驗正解

01. Here's a copy　02. is to confirm　03. Let me know　04. letter dated

公司服裝規定分析

林恩·羅威 – 人資部門
20××年4月11日

目次
目的
現況
分析
建議事項

目的
本報告的目的是彙報本部門對公司服裝規定的檢討結果，並根據結果提出建議。

現況
公司目前要求所有同仁遵守嚴格的正式服裝規定。這是為了營造專業的工作環境，以因應客戶、簽約對象、合作夥伴的頻繁來訪。容許的服裝包括適合正式商務場合的西裝、運動夾克、西裝褲、西裝裙。

分析
業務部門之外的其他部門，見到外來客戶和合作夥伴的頻率不高，同仁們似乎對於穿著正式服裝感到不便，並明確表示更舒適的服裝能夠提升員工士氣，並培養職場上的自主性和創意力。根據20××年4月2日匿名進行的問卷調查結果顯示，大多數同仁偏好在職場上穿著更舒適的服裝，比例達83%。

建議事項
人資部門建議將全體同仁的服裝規定從正式西裝調整為商務休閒。商務休閒包含西裝、褲裝、外套、襯衫、短裙、連身裙。儘管他們不是正式衣著，卻是適合商務環境的著裝。本部門確信同仁能夠判斷服裝是否適合工作環境。如有服裝不當或不專業的同仁，將進行個別處分。

事後檢討報告

環境保護局
美國煉油廠
20✕✕ 年 10 月 21 日

本報告包含針對塔科馬煉油廠火災事件的行動後檢討（After Action Review，AAR）及後續建議。本事件發生期間為於 20✕✕ 年 2 月 13 日至 20✕✕ 年 2 月 16 日，發生地點為位於華盛頓州塔科馬的塔科馬煉油廠。

本行動後檢討包括四大項：
1. 事件概要
2. 事件起因
3. 應變效率
4. 後續事故預防建議事項

1. 事件概要
20✕✕ 年 2 月 13 日週二上午 8 點 50 分左右，在 9D 油罐發現到天然氣外洩跡象。約 20 分鐘後，於 9 點 10 分時火勢迅速爆發。9 點 34 分，天然氣火災控制中心（Gas Fire Control，GFS）接獲尋求協助的聯絡。約 11 點 00 分，GFS 工作人員開始抵達事件現場。因天然氣持續外洩，火勢直到 2 月 16 日 13 點 00 點才得到控制。

2. 事件起因
經過徹底調查後，GFS 調查委員會的結論是由於 9D 油罐受壓過大，致使大量甲烷外洩，導致火災發生。

3. 應變效率
GFS 的應變措施執行狀況良好。他們在事發兩小時內抵達現場，一抵達現場，應變人員便展開現場安全維護，並即刻處理安全問題。然而，在關鍵的最初幾個小時內，現場並沒有足夠的應變人員能做出完善的處置。

4. 事故預防建議事項
a. 環境保護局應針對大型火災事件制定個別的應變作業流程，該應變作業流程應具備確切的應變標準流程及措施。
b. 品質查驗人員應開始針對天然氣外洩，尤其是涉及甲烷外洩，執行定期檢修。

小測驗正解

01. occurs/lasts from　02. report is to　03. have an assertion that　04. dealt with

文書範本 1

會議記錄

企劃部
每月小組會議
20×× 年 9 月 13 日

會議主持人克里斯·塔斯科特於下午 4 點 30 分宣布會議開始。

出席人員
克里斯塔·斯科特（會議主持人）
謝麗爾·沃克
邁克·基弗

缺席人員
（無）

核可會議記錄
20×× 年 8 月 16 日的會議記錄不須修改，通過核可。

動議
動議：會議主持人克里斯塔·斯科特提出 20xx 年 10 月 1 日進行產品測試實施案。
投票：贊成 3、反對 0
決議：通過動議提案。

會議過程
· 會議主持人克里斯塔·斯科特進行每月財務報告。
· 謝麗爾·沃克進行線上宣傳更新報告。
· 邁克·基弗宣佈最近聘用喬治雅·德納姆為新祕書。

結尾
指定下次會議時間為 20×× 年 10 月 15 日下午 4 點 30 分。

由會議主持人克里斯塔·斯科特於下午 6 點 00 分宣佈會議結束。

20××年3月3日行銷策略會議

宣布開會

會議主持人李察·張上午9點35分宣布會議開始。

出席人員

會議主持人李察·張，行銷部經理

瑞亞·海恩斯，業務部經理

約翰·瑪雅，行銷部副理

卡拉·詹寧斯，行銷部襄理

缺席人員

安德烈斯·布朗，業務部副理

動議：

動議：業務部經理瑞亞·海恩斯提出取消5%折扣券。

投票：贊成1、反對3

決議：不通過動議提案。

討論事項

- 約翰·瑪雅對目前正在執行的線上行銷活動進行簡報。

- 討論廣播、電視等的線上廣告替代方案。

- 達成必須提高預算以展開更積極的行銷活動的共識。

結論

出席人員同意在下次會議20××年3月10日確定改變行銷策略之前，進一步探討此案。

休會

因無其他事項，由會議主持人李察·張於上午11點15分宣佈會議結束。

會議記錄由指定記錄人員克拉·約翰遜提交。

小測驗正解

01. Consensus was reached
02. approved without modificaton
03. was designated
04. review the matter

工作交接文件

交接文件

職稱：	生產管理人員
交接日期：	20×× 年 5 月 13 日
交辦人：	約瑟夫·約翰斯通
接管人：	朱迪·思莫斯

職務說明

監督戴爾工廠的飲料生產以滿足客戶需求，並盡力達成成本管理、廢棄物、安全性、生產力及產線效率等相關公司標準。

主要文件

· 20×× 年營運及生產計畫表
· 保健及安全條例說明

主要工作

· 編寫戴爾工廠年度生產計畫表
· 達成品質及安全相關業務執行標準
· 監管設備及裝備的可靠度，盡可能縮短停機時間
· 透過建立員工關係、招聘、培訓及有效溝通，以培養員工能力

近期及目前計畫現況

· 保養 6 號及 7 號產線（進行中）
· 檢討健康及安全規定（20×× 年 5 月 15 日截止）

交接文件

撰寫人：加里·特納
職稱：技術寫作人員
交接日期：20×× 年 7 月 11 日

職責概述：
· 與工程師及產品經理密切合作，編寫綜合使用者及管理人員的軟體文件
· 確保所有技術手冊寫作風格一致，且符合實際產品功能

主管及報告程序：
安東尼·努涅斯先生（每週五 18 點 00 分前以書面提交工作報告）

定期會議、報告或程序：
· 每週一 9 點 00 分進行小組週會

主要文件及參考資料（見本文附件）：
· 內部用語表（20×× 年 6 月 20 日最後更新）

近期及目前計畫進行情況：
· 研發 AIPK 2.0 使用者手冊（截止日期：20×× 年 8 月 31 日）
· 持續更新內部用語表

工作檔案位置（紙本資料及電子檔案）：
公司伺服器的「TW Team」資料夾

通訊錄（內部、外部）

姓名	職稱	電話號碼	電子信箱地址
安東尼·努涅斯	主管	070-383-5844	anunez@mail.com
傑米·黃	組員	070-383-5850	jhwang@mail.com
傑西·貝德福特	組員	070-383-5851	jessieb@mail.com

小測驗正解

01. company standards　02. ensure that　03. taken over
04. Minimizing downtime

文書範本 1

網路釣魚郵件公告

最近北岸會員收到包含附加檔案，或連結到非北岸網站的信件。這些附件多為惡意連結，不可點開。

若信件或訊息要求提供密碼或信用卡資訊，建議忽略並盡快聯絡北岸。北岸的方針是絕對不會要求用戶提供個人資料或帳戶資訊。若該信件要求下列資訊，請務必留意。

- 北岸的帳號和密碼
- 社會安全碼
- 信用卡號
- 信用卡 CCV 碼

當收到可疑信件時，請轉寄至 support@northshore.com 通知北岸客服。

梯馬特賣場暫停營業

本賣場自今日 20XX 年 6 月 15 日晚上 7 點起,將暫停營業進行賣場維修及整備,目前重新開幕的日期未定,感謝各位的協助,造成不便,我們深表歉意。

在我們進一步公告前,歡迎各位前往任一位於您附近的分店購物。

梯馬特賣場 #344
92718 加利福尼亞州爾灣區希爾克雷斯特巷 3349 號

梯馬特賣場 #348
92718 加利福尼亞州爾灣區前景街 1313 號

梯馬特賣場 #351
92718 加利福尼亞州爾灣區戈爾迪巷 3261 號

小測驗正解

01. closed for maintenance   02. apologize for   03. is our policy
04. until further notice

提案計畫書

文書範本 1

商務提案計畫書

對象：位於密西根州紹斯菲爾德市老鷹道 1657 號的梯馬達公司
提案方：位於密西根州喬普林市巴約道 3329 號的斯賓克輪胎公司

大綱
本提案計畫書的目的是讓梯馬達公司和斯賓克輪胎公司建立策略合作關係。

介紹
斯賓克開發優質的汽車輪胎已有超過 100 年的歷史。當貴公司把我們製造的輪胎和業界類似產品相比，便會知道我們具備更好的規格和優勢。

產品和服務簡介
斯賓克專精於製造轎跑車和小轎車。斯賓克的輪胎綜合了運動型外觀、耐用性和全天候摩擦力，尤其適合潮濕和積雪路面。

價格

尺寸	單價（個）
185/60R15	119.45 美金
185/65R15	123.45 美金
185/70R15	127.45 美金

優惠
· 所有輪胎均通過 SDA 認證。
· 6 年 / 60,000 英里的保固。
· 訂購 50 個以上可享 8 折優惠。

歐洲商務旅行提案

對象：基斯莫爾貿易公司的布蘭·登帕克
提案方：ADS 旅行社的蘭德爾·李

公司介紹

ADS 是一家持有營業執照的旅行社，位於德克薩斯州休士頓。我們專為公司、個人和團體提供各式各樣的旅行套組。

商品介紹

我們的商業旅行方案涵蓋歐洲所有國家，包括土耳其和地中海國家。旅行期間最長可達五天，並使用各種交通工具，包括飛機、火車和遊輪。

價格（來回旅行）

艙等	價格（個人）	價格（4 人以上團體）
經濟艙	2,150.00 美金	1,705.00 美金
商務艙	4,225.00 美金	3,805.00 美金
頭等艙	7,400.00 美金	6,555.00 美金

優惠

旅客可至本公司官網上選擇旅行行程表，可選擇的項目有飯店和交通方式等，但不包括娛樂等額外消費。

保險

旅行保險將由旅行社負擔，承保取消行程、遺失行李、緊急醫療，以及旅途中發生的緊急意外狀況。

小測驗正解

01. offer a variety   02. forging a strategic partnership
03. specializes in   04. Compared to similar products

文書範本 1

流程說明書——年度報告

1.0 目的和範圍

1.1 本說明書將說明上呈給邁爾公司執行長的年度報告之準備和提交流程。

2.0 責任

2.1 各任務之主要負責人如下：

2.1.1 營運長（COO）

2.1.2 財務長（CFO）

2.1.3 行銷總監

3.0 定義

3.1 年度報告——每年 1 月向執行長提交列有公司年度銷售額和利潤的綜合財務報告。

4.0 參考

4.1 20×× 年度報告

4.2 年度報告格式

4.3 發貨單記錄本

5.0 程序

5.1. 年度報告由營運長依據執行長的要求準備。

5.2. 年度報告需使用既定的年度報告格式。

5.3. 營運長和行銷總監檢討報告後，最晚須於每年 1 月 31 日前批准。

備註：年度報告包含去年 1 月 1 日至 12 月 31 日之資料。

政策和流程說明書

聖約瑟夫醫療中心
密蘇里州堪薩斯城

政策編號：133
主旨：K-101 的處理和服用方法
目的：向藥劑師提供符合 K-101 藥物處方條件患者的開藥指南。

政策和程序：
I. 符合條件的患者
- 年滿 18 歲。
- 是聖約瑟夫醫療中心的病人
- 持有 K-101 處方。

II. 收到 K-101 處方時的程序：
- 藥劑師通過檢查中央數據庫來驗證處方。
- 藥劑師按照處方分配劑量。
- 藥劑師向病人明確說明如何服用藥物。
- 當原始劑量消耗完後，藥劑師有義務提供患者所需的追加劑量。

III. 處方限制
- 除非主治醫師有建議特別服用劑量，通常病人一天僅限服用 6 錠，每週 42 錠。

作者：埃里克·M·布萊克利博士
生效日期：20××日 3 月 11 日
修訂日期：20×× 年 4 月 27 日，20×× 年 10 月 9 日

小測驗正解

01. prepared by   02. be obligated   03. describes the process   04. In accordance

電子報

凱斯科電子報

20×× 年 3 月號

新上市！

20×× 年 3 月 1 日上市的 INOS 2.0 對 INOS 1.0 做出了重大升級，修復嚴重錯誤並追加全新的功能。INOS 是 Inter-Multitasking Operating System（跨平台多工處理系統）的縮寫，讓使用者可以只用一種方法，管理各種任務的多工系統。請點擊這裡了解更多。

本月最佳員工

三月的最佳員工是軟體研究中心的申瑞秋。過去兩年中，她不斷傾力開發 INOS 2.0，這次 INOS 2.0 能成功上市，她功不可沒。這次的上市預計將於多工處理領域獲得更多關注和投資。

未來活動

3 月 8 日星期三：雲服務系統凱斯科 PRL 上市
3 月 17 日星期五：凱斯科年度開發者大會（凱斯科總部）
3 月 20 日星期一： 尤里卡系統修補程式 4.5 上市

病毒警報！

2 月最後一週惡意病毒肆虐。請各位注意標題為「傑出收據」的信件，此信件中夾帶眾所皆知的可疑檔案，以及標題為「提升耐力新神藥！」的信件，此信件已被證實為有害的釣魚信件。

凱斯科公關團隊
弗雷德里克·克里奇 | 070-3736-1100 | pr@kesco.com

專業人士

- Pro Office 480 用戶每月電子報 -

20XX 年 3 月 11 日

編輯的話

大家好，歡迎來到第五期專業人士，這一期包含關於 Pro Office 雲端功能的最新資訊和建議。

本期目次

- 用 Pro Office 480 做日程管理
- 本月的功能：雲端訊息
- 即將推出的更新
- 推薦欄：IT 分析師馬克斯·布朗

進度管理？簡單吧？那就讓它變得更簡單！

Pro Office 480 的 Work Manager（工作管理人）讓進度管理變得前所未有的簡單。工作管理人是馬蒂·瓊斯投注三年以上心血的發想。它提供能和其他 Pro Office 480 程式相互連動的方便功能。閱讀更多

免費試用！

專業人士為所有註冊會員提供一個月免費試用 Community Help（社區幫助）。當您使用 Pro Office 480 有任何疑問或問題時，你可以透過社區幫助，使用我們 24 小時的線上服務。閱讀更多

取消訂閱

專業人士

信箱：editor@theprofessional.com | 網址：www.theprofessional.com

小測驗正解

01. expected to garner　　02. edition contains　　03. upgrade over　　04. offering a free

文宣

文書範本 1

您的銀行交易安全嗎？
青色銀行
值得信賴的網路銀行交易

青色銀行歷史
19×× 年 創立青色銀行
19×× 年 開辦全國第一家網路銀行系統
19×× 年 進軍中國市場
20×× 年 進軍印度市場
20×× 年 慶祝 40 週年

為什麼您應該選擇我們？
青色銀行是美國最大、最值得信賴的銀行之一。我們專門提供既安全又能防止駭客入侵的網路銀行交易，透過我們的網路和手機銀行系統，您可以安全處理帳單支付、轉帳和資金轉移等服務。

有任何疑問？
可至您附近的青色銀行分行詢問，或至 www.cyanbank.com 獲得更多資訊。

青色銀行總公司
21244 馬里蘭州溫莎磨坊哥倫比亞大道 2807 號

www.cyanbank.com

為什麼您該使用 JTS 寄包裹呢？

JTS 快遞
簡單、快速又可靠

簡單
只需要 3 個簡單的步驟，就能完成包裹寄送，一切都能在線上完成！

可靠
保證運送過程零丟失、無損壞。一旦發生將 100% 賠償！

快速
預計 24 小時內取走您的包裹，並在 24 小時內送達。總共 48 小時！

方便
追蹤包裹從沒這麼簡單過！快上 JTS 快遞的網站查看線上即時追蹤系統吧！

如何開始使用我們的服務？
1. 前往 www.jtsexpress.com
2. 加入會員
3. 下單！

小測驗正解

01. for more information　02. one of the most　03. Check out　04. celebrated

文書範本 1

謝爾曼能源公司
電話留言

收件人：凱莉·金

來電資訊

☑ **先生** ☐ **小姐**　約瑟夫·海

公司：農場交易公司

電話：(562) 344-3562

☑ 請回電　　　　　　　　☐ 將再來電
☐ 回電給您　　　　　　　☐ 要求會面
☑ 留言（如下）

留言內容：
來電確認 9 月 23 日的訂單，希望今日能得到您的回電。

日期：9 月 25 日 20XX 年　　　　　　時間：下午 3:05
留言負責人：帕特·艾姆斯

留言訊息

給 阿朗佐·戈麥斯

日期 20XX 年 7 月 5 日　　　　　　　　　時間 上午／ 下午 11:20

來自 胡安·阿吉拉

電話 (311)-494-5561

☑ 來電找您	☐ 要求回電	
☐ 回電給您	☑ 將再來電	
☑ 預約見面	☐ 緊急	
☐ 發送信件	☐ 發送傳真	

留言

希望盡快見面討論合約條款。下午將再打電話預約會面。

留言人 米絲媞·卡普爾

小測驗正解

01. called for　　02. left a message　　03. called back　　04. make an appointment

 **文書範本 1**

關鍵小工具推出創新平板系列：專業互動

洛杉磯時報 20×× 年 8 月 24 日 – 關鍵小工具今天推出了第一代名為「專業互動」的平板電腦，具有突破性的多點觸控螢幕和第 7 代四核心處理器。

這款平板電腦於洛杉磯科技展覽會由執行長珍妮特·希爾曼介紹亮相。珍妮特·希爾曼說：「這款全新系列的平板電腦，可以說是我們至今最自豪的成就。我相信消費者一定會喜歡這項產品，因為它容易上手、設計獨特。」

專業互動提供下列功能：
· 多點觸控螢幕較前一代的互動 X 靈敏 3 倍
· 第七代四核心處理器
· 重量為 550 克
· 可選擇 128GB 或 256GB 的儲存容量

想了解更多規格、客製選項和配件，請至 www.keywidget.com/interactivepro。

關鍵小工具簡介
關鍵小工具總部位於加利福尼亞州托倫斯，專精於為消費者帶來各種突破性的高科技產品。

記者
勞拉·海耶斯
lauraehayes@keywidget.com
(901) 763-9146

卡爾霍恩物流捐贈 100,000 美元
協助佛羅里達州颶風救災

佛羅里達州邁阿密 –20×× 年 9 月 20 日 – 卡爾霍恩物流已投入現金 100,000 美元及救災物，協助佛羅里達州南部颶風災區的救災工作。

琳達颶風於 9 月 18 日晚上重創該地區，摧毀了數千座房屋，並造成數百人受傷。運營部資深副總裁埃里克‧弗洛雷斯說：「我們將竭盡所能，為那些受災戶提供救濟。」

卡爾霍恩物流有災難時期提供重要援助的悠久歷史。運輸物資一直是卡爾霍恩物流的專業領域，尤其是 20×× 年卡翠娜颶風，卡爾霍恩物流將 10,000 箱食品、藥品和衣物等運送給受災戶，幫上了大忙。

卡爾霍恩物流介紹
卡爾霍恩物流是一間全球物流公司，每天在世界各地運送數百萬個包裹。卡爾霍恩物流在全球 25 個國家皆設有物流倉儲，並雇有 900,000 名員工。若想進一步瞭解卡爾霍恩物流，請至 www.calhoun.com。

聯絡方式
羅納德‧卡茨馬雷克
(542) 484-7130
pr1@calhoun.com

卡拉‧安德伍德
(542) 484-7131
pr2@calhoun.com

小測驗正解

01. was introduced　02. a long history　03. employs　04. offers the following

文書範本 1

SD 產品保固證明

本保證書適用於 SD inkjet 5460 印表機。

保固期間
本保固保證提供自購買日起 12 個月的免費維修。

保固條款和條件
保固期間 12 個月，保固包含任何製造瑕疵，以及因正常使用所產生的所有問題。

若您在保固期間內根據本保證書向 SD 索賠，SD 將
1. 使用品質和功能相同的零件維修產品，或
2. 以相同型號的產品為您更換。若您願意，也可以替換成擁有相同功能的類似產品，或
3. 退換貨時依您購買的價格全額退費給您。

獲得保固服務
可至您附近的服務中心，出示保固卡和購買收據，即可獲得免費保固服務。

責任限制
本保固不保障因疏忽、錯誤使用、濫用、自然災害、異常電壓，或因受潮、污染、雷電等意外情況所造成的損壞。此外，本保固也不包括未經授權而改動、調整或修理所導致的人為損壞。

保固狀態
SD Inkjet 5460 印表機
您的 SD 保固期將於 20×× 年 6 月 4 日到期

序號：4DC9403M2D
產品編號：G4809AT
產品名稱：SD Inkjet 5460 印表機
保固檢查日：20×× 年 9 月 13 日

保固類型：基本保固
服務類型：SD 異地維修支援
狀態：(有效)／已過期
開始日期：20×× 年 6 月 5 日
結束日期：20×× 年 6 月 4 日
服務層級：全球服務、標準材料處理、標準零件、訪問客戶及維修中心取貨

產品更換說明
若您欲更換瑕疵品，必須使用 UPS 或 FedEx 等可追蹤的物流公司，將產品寄送至附近的服務中心。我們在收到瑕疵品後，將為您寄出替換用產品。

請檢視我們的保固政策。

小測驗正解

01. guarantee the provision　02. cover damage　03. ship a replacement
04. warranty applies

文書範本 1

收件人：辦公市場全體員工
發信人：內特·肯德爾，負責人兼執行長
日期：20××年7月11日
主題：公司重組

今天我要宣布公司組織變動的消息，這可能將對我們親愛的辦公市場全體員工帶來莫大影響。雖然這項決定非常艱難，但我相信為了公司未來的生存和發展，這是必要的決定。

由於現今市場的變化和經濟衰退的負面影響，公司必須跟進並適應未來。面對激烈的競爭，我們打算讓辦公市場成為更靈活的公司以滿足客戶的需求。為了達成這些嘗試，未來兩週內將有350名員工離開公司。

感謝這些同仁對公司的貢獻，我保證在這艱難的過渡期中，公司仍會盡最大的努力幫助他們。

各位很快就會從部門主管口中得知，所屬的單位將會發生什麼結構性的變化。如有任何問題，請各位與主管溝通。在此真心感謝各位在這段艱難時期對公司的諒解。

收件人：克里斯托弗·沙利文 , 羅傑布·洛克 , 希爾德·周
副本：基思·傑克遜 , 哈維爾·威廉姆斯
寄件人：珍娜·斯科特
日期：20×× 年 3 月 11 日
標題：5 月 15 日的培訓研討會

由於公司湧入大量新進員工，20×× 年 5 月 15 日全部門將於大酒店舉辦培訓研討會。

研討會將包含公司規則、一般政策、服裝規定和商業禮儀等主題，也會安排勵志演講者演講，和促進團隊合作的小組活動。我希望這場研討會能變成一年一度的活動，因為我相信這能大幅提振員工士氣。

請各位參考附件的培訓研討會議程，歡迎各位給予回饋及評論。

附件：5 月 15 日培訓研討會議程

 文書範本 1

標題：詹姆斯·薩頓，您被邀請加入 EZ 網站 WearPlus.com

詹姆斯·薩頓您好，

推薦您 EZ 的專用購物網站
WearPlus.com
我們將提供各種優惠，讓您享受購物樂趣。

天天特價：設計師品牌最高可享六折優惠，還能找到合適的居家好物。
免運優惠：購物滿 69 美元以上即可免運。
與 EZ 連動：只需以您原有的 EZ 帳號登入即可購物。

每天都有新活動，現在馬上加入吧！

取消訂閱｜幫助

客服專線 1-866-235-5443

標題：在海德威普拉絲獲得七折以上優惠

在網頁瀏覽｜轉寄給朋友

歡迎光臨海德威普拉絲
獨家優惠　完美省錢

馬上抓住獲取優惠的機會。

線上網站

在我們的網站上購物，將節省 30% 以上！

註冊 >

行動 APP

可獲得超值優惠券及更多優惠。

立即下載 >

實體店面

來找經理，他們將會為您找到最適合您的物品。

尋找據點 >

現在就追蹤我們吧。

請勿回覆本信件。若要聯繫我們，請點擊這裡或撥打 1-800-776-5000。

若想更改您的資料設定和訂閱資訊請點擊這裡。若想取消訂閱請點擊這裡。

小測驗正解

01. access to　02. to recommend　03. Follow us online　04. Take advantage

請求信

 文書範本 1

標題：索取「活躍肝臟」樣品

斯賓克先生您好，

我叫李賢貞，我在韓製藥負責行銷。我們是韓國第二大連鎖藥店，且現在仍持續尋找擴大我們產品線的方法。

我們有興趣進口貴公司的一些產品，我們對「活躍肝臟」特別感興趣，據我所知該藥品主打保持健康肝臟。

請問您是否能提供我們免費的「活躍肝臟」樣品呢？提供一份產品或30錠便足夠。此外，若您還能提供詳細資訊，例如成分、服用方法、營養資訊等資料我們將十分感謝。也請您同時提供關於大量訂購的報價。若有獲利性，我們希望能和貴公司簽訂獨家合約。

您可以透過此郵件地址與我聯繫，或以 82-2-990-3933 這支電話與我聯繫。期待能得到貴公司的回覆。

祝好。

李賢貞
行銷經理
韓製藥

標題：推薦信申請

伯曼先生您好，

希望您過得安好，此信目的是想請您幫我個大忙，我要申請凱勒姆的資深行銷一職，希望能請您為我寫推薦信。我和您曾密切合作十幾年，因此我最先想到能幫我寫推薦信的人就是您。加上您信賴度極高，我也總是尊重您的判斷和建議。

為了您方便，我隨信附上推薦信的草稿，供您作為模板使用，同時也附上一份我認為的成就和強項的清單。這份清單一定能喚醒您對我在 20×× 和 20×× 年間所做的專案的記憶。

如果您礙於時間或意願不便幫我寫推薦信，我也完全理解。如果您願意的話，您可以按照您的想法，自由修改附檔中的草稿。我也一起附上我的履歷和招聘公告供您參考。如果您還有任何疑問或想知道更多資訊，請撥打 324-339-4554 與我聯繫。

感謝您的諒解。

祝好。

艾米·利奧

小測驗正解

01. send us　　02. look forward　　03. in charge　　04. be grateful

提案計畫信

 文書範本 1

標題：「活躍肝臟」合作提案

斯賓克先生您好，

我是韓製藥的李賢貞，感謝您寄給敝公司要求的「活躍肝臟」樣品，我們也在一週前收到狀態良好的產品。

我們在進行樣品研究及大範圍的市場調查後，確信「活躍肝臟」在韓國的市場極具潛力，因此我們希望能和貴公司簽訂獨家進口的合約。若您們允許敝公司在兩年間獨家進口和販售該產品，我們也將提供以下優惠。

· 讓「活躍肝臟」在各通路廣為流通及宣傳
· 保證每年至少 1,500 份的銷售量
· 考慮將來進口貴公司的其他產品

請讓我知道您對這項提案的想法，您可以寄電子郵件給我，或撥打 82-2-990-3933 與我聯繫。

祝好。

李賢貞
行銷經理
韓製藥

標題：會議提案

嗨，麥克，

我是人資部門的麗茲。最近客戶端的安全漏洞問題解決得並不順利，且需要馬上著手處理此問題。

因此我提議我們（人資部）、公關部、技術部，以及業務部能一起開會。我認為你是業務部中最了解這件事的人，所以希望你能親自參與這場會議。

如果上述提案可以的話，能告訴我你明天方便的時間嗎？

麗茲

詢問信

 文書範本 1

標題：詢問火花塞電線

詹寧斯女士您好，

我是約翰·哈，是 TK 汽車廠的生產經理，目前正在尋找火花塞電線的供應商。

我想知道貴公司是否有火花塞電線，如果有的話，是否能提供報價。我尤其想知道貴公司有哪些種類的電線，每種類型的價格為何。我也想詢問延長保固期及付款方式。

請務必回覆所有相關資訊和選項，謝謝。

祝好。

約翰·哈
生產經理
TK 汽車廠
070-3651-9302
www.tkmotors.com

標題：卡樂廣告線上服務

您好，

我是琳達·巴頓，在卡樂廣告服務。卡樂廣告是提供數位行銷服務的領先供應商之一。我在尋找可能需要我們服務的公司時，偶然發現了貴公司。

貴公司目前看來已經擁有極高的線上認知度，因此我想進一步了解貴公司是否有意願使用我們有效的線上策略，來鞏固貴公司的知名度呢？若貴公司願意告知我們今年的目標，我非常樂意說明我們如何藉由線上廣告，協助貴公司達成該目標。

若貴公司想和我們討論合作的可能性，請隨時寄信至 linda.batten@collor.com，或撥打 (353) 477-9004 與我聯繫。期待能與貴公司合作。

祝好。

琳達·巴頓
卡樂廣告
www.color.com

小測驗正解

01. revert back　02. online presence　03. are looking　04. interested to know

文書範本 1

標題：內特費爾特服務中止

布萊恩·海登您好，

我們在 3 月份寫信通知您，我們將停止提供以下服務：

· 內特費爾特翻譯
· 內特費爾特圖庫
· 內特費爾特熱門職位

上述服務將自 20×× 年 8 月 11 日起終止，目的是為了簡化我們的服務內容，以便將來為您帶來更好、更有效率的服務。

這些服務已從 www.netvert.com 的首頁移除，因此您不能再使用它們。不過請您放心，我們已開始著手研發更適合的替代服務。

若造成您的任何不便，我們深表歉意，感謝您的協助和諒解。如果您還有任何問題，請至客服中心網址 www.netvert.com/help。

請勿回覆本信件，此信箱將不會做出任何回覆。

標題：異常的信用卡活動

收信人：溫迪‧帕拉
帳號 4474609771
信用卡號 8330-1833-1731-9012
日期：7/30/20××

親愛的顧客您好，

我們發現 20×× 年 7 月 30 日您的亞利桑納銀行信用卡有異常消費，為了保護您的資訊，您必須先進行驗證，才能繼續使用此卡。

若您要查看和驗證此筆消費，請上 www.bankofarizona.com/protection 或立即撥打 1-800-832-9412 與我們聯繫。我們將在您驗證後，取消此卡的任何限制。若您已經驗，請忽略本通知。

由於這並非安全的溝通管道，因此請勿回覆本信件。若您對本人帳戶有任何疑問或需要協助，請撥打帳單上的電話或點擊「聯繫我們」的網址 www.bankofarizona.com。

小測驗正解

01. your protection   02. be assured   03. have any questions
04. disregard this notice

交易信

標題：Wireless.com 訂單確認

訂單確認
訂單編號 #901-3846119-391

大衛·李您好，
感謝您訂購我們的產品。您於 10 月 9 日的消費訂單詳情如下。您的交易明細可在訂單發票上找到。如果您想查看訂單狀態，或對其進行任何更改，請至 Wireless.com 查看「我的訂單」。

預計送達日：20×× 年 10 月 11 日 星期二
寄送方式：一般貨運
寄送地址：大衛·李
　　　　　49855 密西根州馬奎特鐵道街 4467 號

訂單資訊：
V-MODA 耳機　　RS 210
商品小計：　　　$58.95
運費 & 處理費：　$2.50
總計：　　　　　$61.45

感謝您使用 Wireless.com 購物。點擊這裡可瀏覽更多商品並獲得紅利點數。

標題：彈性部落客密碼重設

彈性部落客

哈囉，傑西卡，

我們已收到您要求重設彈性部落客密碼的請求。若您提出此請求，請點擊下方連結。

重設密碼

此連結有效時間為 24 小時或直到您重設密碼。

若您未要求重設密碼，請忽略本信件。很可能是其他用戶誤用了您的用戶名稱。除非您點擊連結，否則您的帳戶將不會進行任何更改。

彈性部落客團隊

請追蹤我們的臉書和推特。

小測驗正解

01. status of your order   02. click on   03. can be found   04. reset your password

投訴信

文書範本 1

標題：牛奶訂單的有效期限

摩爾先生您好，

我是杏仁烘培坊的卡拉·斯奈德。昨天我們收到向您訂購的 120 箱牛奶，但是發現其中有 40 箱牛奶已過期。

我們在 8 月 21 日收到您的牛奶，其中的 80 箱正常，但其餘 40 箱的有效期為 8 月 19 日。由於我們的烘培坊自詡只使用最新鮮的材料，因此我們無法使用過期的 40 箱牛奶，而這也導致我們熱銷的迷你瑪芬生產延宕和產量不足，損失至少達 5,000 美元。

我們希望能收到您對這起事件的回覆，以及將如何補償我們的損失。請盡快撥打 (998) 867-0987 或寄信至 ksnyder@gmail.com 與我聯繫。

祝好。

卡拉·斯奈德
杏仁烘培坊

標題：121 號班機登機證

內華達航空公司您好，

我原本應該搭乘 20××年 4 月 21 日內華達航空 121 號班機飛往巴黎，但不幸地辦理登機時，地勤給了我錯誤的登機證。

直到我都快走到錯誤的登機口才發現，導致我無法及時趕到正確的登機口登機。這項失誤讓我進退兩難，直到兩天後我才能訂到新的機票。

此事件帶給我很大的精神和經濟壓力，我希望貴公司能付出適當的補償和行動，以免同樣的事情再次發生。希望能盡快得到貴公司的回覆。

祝好。

克里斯蒂娜·格里芬

小測驗正解

01. in a difficult situation　02. contact me　03. caused a delay
04. mistakenly given

邀請信

文書範本 1

標題：邀請您參加年度股東大會

敬愛的林卡斯特股東，

誠摯邀請您參加將於美東時間 20×× 年 2 月 28 日星期二早上 11 點，於佛羅里達州奧蘭多的奧蘭多會議中心，舉行的林卡斯特年度股東大會。詳情請見附檔資料。

今年的會議議程如下：

1. 批准關於 20×× 年 3 月 24 日召開的 20×× 年度股東大會的會議記錄。
2. 審核並通過公司 20×× 年度資產明細表。
3. 任命一位審計員和其報酬。
4. 審議其他業務事項。

我們懇請您出席會議，因為今年的會議將做出重要決定。若您因故無法參加，請派代理人出席。

期待您的光臨。

祝好。

芭芭拉·G·倫特羅
執行長
林卡斯特

標題：誠摯邀請您參加 S&M 報告大會

普森先生您好，

感謝您加入中小企業協會。我們很高興能邀請您參加關於中小企業 20×× 前景的報告大會。

活動詳情如下：

日期：20×× 年 1 月 6 日星期六
時間：下午 2:00 到 4:00
地點：羅尤爾酒店 2 樓
　　　90071 加利福尼亞州洛杉磯南方巷 4478 號
講者：胡里奧·約曼斯博士
主題：20×× 中小企業展望

我們相信這次的報告大會，能提供您對未來一年事業一些有用的想法。隨信附上活動的位置圖，若您確定出席，請於 20×× 年 1 月 4 日前回信。若您有任何問題，可寄信至 help@smba.org 聯絡本協會。

祝好。

布倫娜·布萊爾
主席
S&M 商業協會

小測驗正解

01. will be held　　02. ask for your presence　　03. are enclosed
04. confirm your attendance

文書範本 1

標題：搶先需要您的反饋！

客戶問卷調查

凱爾您好，

搶先一直致力於尋找改善我們產品和服務的方法。若您能參與本次的調查，我們將會非常感謝。您的反饋將提供我們寶貴的資訊，也有助於我們能更好的滿足您的需求。

本問卷約需 10 分鐘。若完成此份問卷，搶先將轉入 1.00 美元至您的帳戶作為感謝。

請點擊下方按鈕開始填寫問卷。

開始問卷

若您在填寫問卷時遇到任何疑問或技術性問題，請寄信至 survey@headstart.com。

在此先感謝您的參與，我們期待得到您的反饋。

標題：分享您對銷售經驗的看法

清晰市場

阿曼達·費舍爾您好，

感謝您成為清晰市場的寶貴賣家。我們的記錄顯示您上週在清晰市場售出 8 件商品，因此我們想邀請您參與一項簡短的調查，分享您在清晰市場上的銷售體驗。您的意見將有助於清晰市場成為對買家及賣家而言，更好的地方。

請點擊下方連結填寫問卷，只需 5 分鐘！或您也可以將下方連結複製貼上於您的瀏覽器。

https://survey.clearmarket.com/survey/seller/clr88301?co+us&smpl_gst=146827313

感謝您的反饋，也感謝您對清晰市場的支持！

祝好。

清晰市場客戶體驗團隊

小測驗正解

01. short survey　　02. advance　　03. meet your needs　　04. express your opinions

感謝信

 文書範本 1

標題：感謝您的主題演講

嗨，克雷格，

感謝您上星期五於百忙之中到我們公司演講。您選擇的主題「強大溝通力的力量」至今仍讓我們深感共鳴。

您的演講可以說是兼具教育意義和娛樂性，帶給我們許多員工鼓勵和啟發，讓他們也想磨練溝通技巧。此外您準備的圖片和補充資料，也幫助我們更理解您所要傳達的訊息。我們衷心感謝您所做的努力。

希望很快我們能再次合作，也恭喜您這次出色的表現。

桑德拉·科爾格羅夫
人事經理
卡爾霍恩物流

 **文書範本 2**

標題：感謝給予面試機會

格林先生您好，

感謝您今日抽空和我面試，我很高興見到您，以及了解更多關於心智科技工程師職位的資訊。

和您面試後讓我更想成為心智科技的一員，我相信我在 PXO 的實習經歷，能幫助我更有效率地執行這份工作。

如果您還需要我提供更多資訊，請隨時與我聯繫。期待能收到您的回覆，再次感謝您的好意。

祝好。

卡羅琳·芬克

小測驗正解

01. appreciate your efforts   02. say the least   03. can work with you
04. your busy schedule

買賣合約

文書範本 1

物品買賣合約

本買賣合約（以下稱「合約」）自 20×× 年 2 月 14 日起生效，雙方為公司位於德克薩斯州休士頓鳥泉巷 641 號的史密斯公司（以下稱「賣方」）和公司位於芝加哥伊利諾州點街 1450 號的貝納維德斯公司（以下稱「買方」）。

購買商品：
買賣雙方同意根據本合約之條款和條件購買且出售以下產品。

產品描述	數量	單價	總價
史密斯 TY189 晶片組	200	$55.00	$11,000.00

支付方式：
當賣方交付本合約所指定之貨物時，買方應以現金支付賣方 11,000.00 美元。

送貨：
賣方將安排由買方選擇的物流業者運送貨物，並應於 20×× 年 2 月 28 日到貨。

保固：
賣方保證貨物在原料或製作技術上無任何瑕疵。

本合約由史密斯公司的業務代表艾倫·克蘭，和貝納維德斯公司的財務總監拉里·M·吉列梅特代表簽署。

賣方：
史密斯公司
艾倫·克蘭
艾倫·克蘭
業務代表

買方：
貝納維德斯公司
拉里·M·吉列梅特
拉里·M·吉列梅特
財務總監

事業買賣合約

本合約於 20×× 年 5 月 17 日，由 SD 集團（以下稱「賣方」），總公司位於韓國首爾市麻浦區大井路 75 號，以及 TK 集團（以下稱「買方」），總公司位於韓國首爾鐘路區社稷路 2 街 48 號所簽訂。

1. 事業資訊
本企業包含下列資產：根據本合約，賣方應將庫存、原料、成品轉移給買方。

2. 購買單價和款項支付方式
包括所有銷售稅在內，買方應以保付支票支付賣方 14,200,000.00 美元。

賣方保證 (1) 賣方擁有事業的法定所有權，以及 (2) 保證賣方具有出售和轉讓事業權的所有權利。

買方具有檢查與事業相關財產的機會，並以原有條件接受該事業。本約以大韓民國法律為準擬定，並受制於該法。

20×× 年 5 月 17 日由以下代表人簽署、蓋章並交付。

賣方　　　　　　　　　　　　　　買方
SD 集團　　　　　　　　　　　　　TK 集團

　金浩林　　　　　　　　　　　　*玉英敏*

小測驗正解

01. agreement is made　　02. by and between　　03. arrange for delivery
04. made effective

253

聘雇合約

文書範本 1

聘僱合約

20××年2月14日洛克林公司（以下稱「雇主」）和魯迪‧L‧亞當斯（以下稱「受雇人」）簽訂如下聘僱合約（以下稱「合約」）。

1. 職務
雇主自20××年3月1日起聘請受雇人為銷售人員，至聘僱關係結束為止。受雇人同意將盡力為雇主銷售產品，且竭盡所能發揮自身能力和經驗，誠實且勤奮地工作。工作時間為一週40小時。

2. 薪資
雇主將按照支付薪水的程序，支付受雇人年薪30,000美元作為銷售工作之報酬。

3. 保密
受雇人同意不得在未獲得雇主書面同意前，將公司機密洩漏、公開、轉達給第三方。違約時，雇主有權採取包括要求損失賠償等相應之法律措施。

4. 福利
受雇人享有每年21天的有薪假和5天病假。

5. 期限及合同終止
雇主和受雇人需在一個月前書面通知對方，方可解除本合約。

勞拉‧斯旺
勞拉斯旺，人資經理
洛克林公司（雇主）

魯迪‧L‧亞當斯
魯迪‧L‧亞當斯
（受雇人）

拱頂石營造
聘僱合約

本合約由拱頂石營造公司（以下簡稱「公司」）與約翰·諾里斯（以下簡稱「決策者」）於 20×× 年 7 月 1 日簽訂。

職務和雇用範圍

自 20×× 年 8 月 1 日起，決策者將擔任公司負責人兼執行長，並接受公司董事會指示工作。僱用期間，決策者將對公司付出一切努力和時間，並盡最大的能力公平且正直地履行本條約所明示之決策者義務。

合約終止

決策者和公司任一方皆同意無論是否有正當或特殊理由，只需以書面通知另一方，隨時皆可終止僱用關係。但決策者有權依終止雇用關係時的情況，獲得遣散費和其他福利。

合約期間

本合約自生效日期起為期 4 年。且最晚在本合約期滿前 90 天，公司和決策者需討論是否續約，以及在什麼條件下續約。

薪資

自生效日起至 20×× 年 6 月 30 日止，公司每年需支付決策者 400,000 美元的薪資，作為此期間決策者的工作酬勞。

小測驗正解

01. annual salary　　02. hereby employs　　03. duties faithfully
04. terminate employment

 **文書範本 1**

著作授權合約

本著作之授權合約（以下稱「合約」）將由藝術科技（以下稱「授權人」）和梅森公司（以下稱「被授權人」）於 20×× 年 5 月 15 日（以下稱「生效日」）簽訂。

准予授權

被授權人在授權範圍內，不得壟斷或轉讓被授與利用著作之權利。被授權人可在其事業上使用授權人之著作，且可複製、製作、販售，此外不得使用於其他目的。唯獲得授權人書面許可時，被授權人才得以將著作物用於其他用途。

條款和終止

本合約將自生效日起生效，具有 2 年效力，且可自動續約 1 年，除非任何一方最晚於每年合約到期日 60 天以前，以書面通知另一方不再續約。

版稅

被授權人應每月支付授權人作品總銷售額的 5% 作為版稅，且應於每月月底前的 10 天內支付。

下列簽署人將於上方明示的第一天起執行本合約，以茲證明。

授權人	被授權人
藝術科技	梅森公司
_____	_____

最終用戶許可協議

通過安裝、複製、連接或以其他方式使用 ITSS 軟體 2.1（以下稱「ITSS 軟體」），您（以下稱「最終用戶」）同意接受本最終用戶許可協議（以下稱「EULA」）的條款約束。若您不同意以下條款，請勿安裝、複製、連接或以其他方式使用 ITSS 軟體。

1. 協議：
本最終用戶許可協議是最終用戶與悉德科技公司（以下稱「悉德科技）之間的法律協議。悉德科技擁有 ITSS 軟體的所有權利，包含其所有權、權益、相關資料、檔案及智慧財產權等。

2. 軟體：
ITSS 軟體包含悉德科技提供給最終使用者使用之相關媒體、印刷品、電子文件、軟體更新、附加零件、網路服務、附件。

3. 權利授予：
ITSS 軟體是授權給最終使用者，而非出售。最終使用者被授權使用 ITSS 軟體的內容如下：悉德科技授予最終使用者以非商業目的使用 ITSS 軟體的已註冊副本，且本權限不得轉讓亦不可壟斷。

4. 終止：
悉德科技保留隨時終止您的授權以及終止您 ITSS 軟體的使用權。若您未能遵守本最終用戶許可協議之任何條款，授權將自動終止，您亦可隨時終止這份授權協議。若您欲終止 ITSS 軟體的使用權，您必須將原因告知悉德科技。在本協議終止時，您同意將您擁有的 ITSS 軟體和所有文件的副本銷毀。

小測驗正解

01. automatically renew　　02. commence as　　03. are authorized
04. reserves the right

保密合約

文書範本 1

保密合約

本保密合約（以下稱「合約」）由天網公司（以下稱「披露方」）和泰爾佩特公司（以下稱「收受方」）簽訂並自下文中指定之日期起生效。

1. 機密資訊
披露方提供之機密資訊、獨家資訊、企業機密指的是標註「機密」、「專利」，或以類似標示之資訊。

2. 收受方之義務
收受方應以保護自身機密資訊和專利資訊的相同層級，來維護披露方的機密資訊。收受方亦不會向員工或任何第三方洩漏披露方之任何機密資訊。

3. 保密期間
若收受方自獲得機密資訊日起超過 3 年，披露方則不得因收受方洩漏披露方之機密資訊，而向收受方索賠任何因違約或盜用商業機密產生的損失。

本合約和各當事人的義務均對各方代理人、指名人和繼承人具有約束力。各當事人均透過其授權之代理人簽署本合約。

揭露方　　　　　　　　　　　收受方

簽名＿＿＿＿＿＿＿＿　　　　簽名＿＿＿＿＿＿＿＿

姓名＿＿＿＿＿＿＿＿　　　　姓名＿＿＿＿＿＿＿＿

日期＿＿＿＿＿＿＿＿　　　　日期＿＿＿＿＿＿＿＿

計畫保密合約

簽署本合約之企業，即貴公司，參與了一項代號翡翠的高機密肯恩科技開發案。本肯恩科技計畫保密合約將陳述要求事項，以保護本計畫之機密。

1. 貴公司不得將項目相關之任何機密資訊，洩漏給未經肯恩科技明確准許觸及此類資訊的員工及任何外部人員。

2. 貴公司同意在收到項目機密資訊前，要求任何獲准接觸資訊的人員在保密文件上簽名。

3. 貴公司同意，允許肯恩科技藉由查帳貴公司的紀錄和資訊設備，以及與貴公司員工面談，來驗證貴公司是否遵守本合約和相關合約。

4. 若貴公司違約，肯恩科技將有權向貴公司追討於 20×× 年 9 月 22 日雙方談妥之違約金。

保安要求：

1. 組成保安團隊，並分配專業的保安管理人
2. 以代號稱呼肯恩科技
3. 以代號指稱保密計畫
4. 定期實施保安教育訓練
5. 任何未經批准的揭露、竊盜或機密資訊遺失皆需通報肯恩科技，以及採取相應的法律措施和解決方案。

小測驗正解

01. code name　　02. take appropriate　　03. protect the secrecy
04. shall be binding

租賃契約

不動產租賃契約

本租賃契約由燭台股份有限公司（以下稱「出租人」）和泰勒能源公司（以下稱「承租人」）於 20×× 年 1 月 23 日簽訂。

不動產
出租人將以本契約明訂之租金，將位於首爾麻浦區世界盃路 78 號的成吉大樓 4 樓出租給承租人。

租賃期間
租期自 20×× 年 1 月 25 日起，至 20×× 年 1 月 24 日止。本契約可自動續約 2 年，除非任一方至少於合約期滿，或續約期限結束前 30 天以書面通知終止合約。

租金
承租人應於每月 5 日先支付 10,000,000 韓元的房租給出租人。

押金
簽訂本契約時，承租人應先支付 200,000,000 韓元的押金給出租人，且出租人應於租約結束時將押金退還給承租人。

維護費和公共費用
出租人有責任於租約期間維持房產的良好狀態，而承租人應負擔一切相關公共費用。

出租人　　　　　　　　　　　　承租人
燭台股份有限公司　　　　　　　泰勒能源公司

＿＿＿＿＿＿＿＿＿＿＿　　　　＿＿＿＿＿＿＿＿＿＿＿

設備租賃契約

本設備租賃契約（以下稱「契約」）由 恩綽奈特（以下稱「出租人」）和 科技模型網（以下「承租人」）於 20×× 年 4 月 15 日簽訂。（統稱為「雙方」）

1. 設備：出租人將以下設備出租給承租人：
兩台影印機（型號：Xerox WorkCentre 7970）（「設備」）

2. 租期：租期自 20×× 年 4 月 20 日起，至 20×× 年 4 月 19 日止。（「租期」）

3. 租金：承租人同意於每月 1 日向出租人支付 250.00 美元之設備租金（「租金」），地址為路易斯安那州梅泰里樹爾斗瑪道 3392 號。若逾期未繳本契約明訂之租金達 10 天以上，承租人同意支付 100 美元的滯納金。

4. 押金：在接管設備前，承租人應向出租人支付 8,000.00 美元的押金，作為承租人履行本契約的保證，以及租賃期間設備因承租人或承租人的代理人發生任何損傷時的保障。

5. 設備的接管與轉讓：承租人有權在租期的第一天接管設備。在租約期滿時，承租人應以當初簽約時同樣良好且正常的運作的狀態，將設備歸還給出租人或出租人的代理人。

出租人 　　　　　　　　　　　　承租人
恩綽奈特 　　　　　　　　　　　科技模型網

_____　　　　　　　_____

小測驗正解

01. take possession　　02. will begin/start on　　03. be responsible
04. at the expiration

合作備忘錄

本合作備忘錄（以下稱「MOU」）由通信多媒體部代表的馬來西亞政府（以下稱「政府」）和 PT 奈特所馬來西亞（以下稱「奈特所」）（統稱為「雙方」）簽訂於 20×× 年 3 月 7 日。

目的
本合作備忘錄旨在提供馬來西亞政府與奈特所之間，資訊通信技術（以下稱「ICT」）的未來發展合作框架。

授權和合作關係
雙方確認彼此的意向，以及完成正當的法定程序後，最晚必須於 20×× 年 10 月 31 日以前簽訂具法律效力的合約，將奈特所軟體提供給政府所有機關部門使用。奈特所將持續支援政府確定的資訊通信技術項目，這些項目將改善馬來西亞的資訊技術取得及軟體產業發展。

合約期間和終止
本合作備忘錄將自簽署日起生效，且效力維持至 (i) 根據本署授權許可生效日、(ii) 20×× 年 10 月 31 日。

非專屬和保密
本合作備忘錄雖非專屬，但雙方必須對本合作備忘錄之條款保密。

雙方代表將於本合作備忘錄署名，以茲證明。

馬來西亞政府　　　　　　　　　　　　PT 奈特所馬來西亞

_____　　　　_____

活動備忘錄

本活動備忘錄（以下稱「備忘錄」）簽訂於 20×× 年 3 月 10 日。

由 ADT 公司（以下稱「ADT」）和 FT 韓國（以下稱「FT」）和波士頓創意中心（以下稱「中心」）共同簽訂。（三方統稱為「當事人」）

目的和義務
當事人同意於 20×× 年 5 月 15 日至 20 日在馬薩諸塞州波士頓共同舉辦 20×× 年創新杯（以下稱「比賽」）。當事人同意本備忘錄並未在他們之間建立任何合約關係，但同意秉持合作精神共同努力，確保活動成功。

合作
比賽的職責和服務包括但不限於：

a. ADT 提供的服務包括：參賽者報名等所有活動相關準備。
b. FT 提供的服務包括：提供所有設備和與活動相關的宣傳。
c. 中心提供的服務包括：提供場地及包含志工在內的人力。

資源
ADT 同意支援活動所需的資金和物資。

期間
根據本備忘錄，當事人所協議之事項將自簽署日起生效，直至活動結束為止。

小測驗正解

01. provides the framework　02. agree to work　03. Following completion
04. rendered

意向書

文書範本 1

購入房產意向書

致圖表公司，

本意向書（以下稱「LOI」）概述卡斯凱爾公司（以下稱「買方」）向圖表公司（以下稱「賣方」）購入以下不動產之一般條款。

1. 不動產
「高塔」（「不動產」）位於康乃狄克州威爾頓學校街 1486 號，由約 12,000 平方英尺的倉庫和辦公空間組成。

2. 購入價格
交易結束時，該不動產的購買價格為 15,000,000 美元（「購買價格」），且以現金支付。

3. 盡職調查
買方應在本意向書簽訂後的 30 天內對不動產盡職調查，包括實物調查，以簽訂具法律約束力的最終買賣合約。

4. 驗收
若接受本意向書的條款，請在下方簽名表示同意，並將一份意向書正本歸還給簽名者。

祝好。

買方：

_____

於 20_____年_____月_____日同意且接受本意向書。

賣方：

_____

興建工廠意向書

肯德爾先生您好，

此為維爾齊公司有意和洛夫塔營造，締結忠清南道新工廠建設的合約意向書。此外，我們希望您在完成合約資料和確定最終方案價格前開始作業。

任何情況下，您不得從事和承諾，超過本意向書中提及之超過總額 5,000,000 美元的工作和承諾。根據本意向書，您需執行的工作範圍如下：

· 現場佈局和調查
· 取得動工所需的許可證和保險
· 動員和安裝臨時建築辦公室

本意向書將作為維爾齊公司和洛夫塔營造之間的臨時合約，直到雙方決定簽訂正式的 CCDC-2 合約，或雙方皆同意中斷本項目的合作。

祝好。

維爾齊公司

———————————————————

於 20×× 年 3 月 1 日同意上述所有條款。

洛夫塔營造

———————————————————

小測驗正解

01. award the contract　02. are acceptable　03. outlines the general terms
04. your acceptance

文書範本 1

合作協議

本協議由波薩德通道與哈博航空（以下統稱為「雙方」）於 20×× 年 7 月 6 日簽訂。

合作名稱和目的
雙方以「波薩德哈博網路服務」為名義建立合作關係，為旅客提供線上購買哈博航空機票的服務。

合作期間
合作期間為 20×× 年 7 月 10 日起，至任一方終止合作關係為止。

貢獻
各方需貢獻以下內容：

波薩德通道：網站相關服務和線上宣傳。

哈博航空：旅行服務、旅行時間表、機票套組和客戶服務。

出資
各方皆應出資合作協議，初始出資為 10,000.00 美元，且各方應按 1:1 的比例分潤或分攤損失。

退出和解散
雙方在此保留隨時退出合作之權利，可以多數決決定結束合作關係，且清償債務後的所有資金將以 1:1 進行分配。

雙方於下方寫下初始提及日期並簽章，以茲證明。

簽名＿＿＿＿＿＿＿＿＿＿ 日期＿＿＿＿＿＿
簽名＿＿＿＿＿＿＿＿＿＿ 日期＿＿＿＿＿＿

加盟合約書

本加盟合約書將根據如下理解和協議,由萊文咖啡(以下稱「加盟總部」)和羅伊·張(以下稱「加盟主」)於 20×× 年 8 月 19 日簽訂。

1. 加盟經營權的授予
加盟總部授予加盟主於京畿道安山市檀園區草芝路 56 號經營萊文咖啡,得以使用加盟總部商標和專有的創立及經營方式之使用權。

2. 加盟金
加盟主同意向加盟總部支付 20,000,000 韓元的加盟金。

3. 權利金
在合約期間內,加盟主同意每月支付加盟總部占總銷售額 8% 的權利金。

4. 加盟期間
加盟期間自本合約完全執行日起至 2 年後結束。任一方皆可提前 30 日通知對方後終止合約。

5. 違約和解約
若發生下列任一情形時,加盟總部有權解約。a. 放棄、b. 破產、C. 犯罪、d. 未能付款、e. 濫用商標、f. 未經授權洩密、g. 反覆違規。

本合約由達雷爾·斯凱曼代表加盟總部簽署,羅伊·張代表加盟主簽署。

加盟總部
達雷爾·斯凱曼

加盟主
羅伊·張

小測驗正解

01. reserves the right　02. operating profits　03. pay a monthly royalty
04. shall contribute

商業文件

 文書範本 1 估價單 Estimate

估價單

日期：20××9 月 24 日
估價單號：48131311
有效期限：20×× 年 9 月 30 日
客戶編號：PI719

地址：派納普公司
92618 加利福尼亞州爾灣市自由大道 2303 號
714-422-0076

數量	描述	單價	總額
20	鐵馬 A4 80 gsm 辦公用紙（500 張）	$9.50	$190.00
2	基本牌訂書機（附 1,000 個訂書針）	$4.50	$9.00
		小計	$　199.00
		增值稅率	%　7.50
		增值稅	$　14.90
		總計	$　213.90

以上並非發票，僅為上述服務／商品的估價。付款將於提供本報價單的服務／商品前收取。

若接受此報價單，請於此處簽名並回傳：＿＿＿＿＿＿＿＿＿＿＿＿

感謝您的購買！
若您對此報價單有任何疑問，請聯繫
羅納德·A·亨斯利（530-773-9081）
【鐵馬股份有限公司】【96001 加利福尼亞州雷丁市拜爾斯巷 2484 號】
電話【530-773-9069】傳真【071-114-4932】【info@tiemapaper.com】【www.tiema.com】

銷售發票

日期：20×× 日 10 月 2 日
發票編號：INV-2013100207
客戶編號：PI719
採購訂單編號：4819
付款截止日期：20×× 年 10 月 24 日

請款單位：

派納普公司
92618 加利福尼亞州爾灣市
自由大道 2303 號
714-422-0076

交貨地址（其他情況）：

派納普公司
92618 加利福尼亞州爾灣市
自由大道 2306 號
714-422-0081

業務負責人：羅納德·A·亨斯利　　　　　　　　送貨日期：20xx 年 10 月 4 日

品項	描述	數量	單價	總額
38601	鐵馬 A4 80 gsm 辦公用紙（500 張）	20	$9.50	$190.00
11381	基本牌訂書機（附 1,000 個訂書針）	2	$4.50	$9.00

小計	$ 199.00
增值稅稅率	$ 7.50
增值稅	$ 14.90
運費及處理費	$ 5.00
折扣	$ -2.50
總計	$ 213.90

請將所有支票開給鐵馬公司

感謝您的購買！
若您對此發票有任何疑問，請聯繫
羅納德·A·亨斯利（530-773-9081）
【鐵馬股份有限公司】 【6001 加利福尼亞州雷丁市拜爾斯巷 2484 號】
電話【530-773-9069】傳真【071-114-4932】 【info@tiemapaper.com】 【www.tiema.com】

斯邁特股份有限公司　　　　　　　　稅務發票

32789 佛羅里達州溫特帕克市泰瑞巷 3774 號
電話：(321) 303-5484，傳真：(321) 303-5480
電子郵件：tax@smiant.com，網站：www.smiant.com

發行日期：20××/10/20
發票編號：200003
採購編號：05-BC-39138393

請款單位：
普拉多健康中心
94520 加利福尼亞州康科德碧曲街 4746 號

產品編號	描述	數量	單位	單價	總額
P48602	野莓	50	箱	$25.00	$ 1,250.00

小計	$　　1,250.00
稅率（10%）	$　　　125.00
合計金額	$　　1,375.00
運費	$　　　　-
支付金額	$　　　　-
未付金額	$　　1,375.00

帳戶資訊
匯款對象：
銀行名稱：東西銀行
帳戶名稱：斯邁特股份有限公司
帳號：302597381981
支付期限：發票設立後 20 天內

銷售收據

斯邁特股份有限公司

32789 佛羅里達州溫特帕克泰瑞巷 3774 號
電話：(321) 303-5484，傳真：(321) 303-5480
電子郵件：tax@smiant.com，網站：www.smiant.com

發行日期：20××/10/30
收據編號：98171
採購編號：05-BC-39138393

購買單位：
普拉多健康中心
94520 加利福尼亞州康科德碧曲街 4746 號

產品編號	描述	數量	單位	單價	總額
P48602	野莓	50	箱	$25.00	$1,250.00

小計	$ 1,250.00
稅率（10%）	$ 125.00
總計	$ 1,375.00

收到金額：1,375.00
付款方式：現金
銷售負責人：托馬斯·瓦萊

感謝您的購買！

貿易文件

 文書範本 1 採購訂單 Purchase Order

有線網計算公司
美國密西根州馬斯基根蜜蜂街 2338 巷
電話：(515) 910-7311 傳真：(515) 910-7350
電子郵件：saunders@wirenet.com 網站：www.wirenet.com

採購訂單

編號：PO-03-187311　　　　　　　　　　　　日期：20×× 年 3 月 3 日

收件方：拉斯克實業公司
　　　　澳洲艾希伯格市魏堡街 79 號

發貨日期	運輸條款	付款條件	貨幣代碼
2017 年 3 月 13 日	CIF	60 天	USD（美金）

產品編號	描述	數量	單位	單價	總額
1413	滑鼠矽膠帽（藍色）	5,500	個	4.50	24,750.00
	（含增值稅）			總計	**24,750.00**

備註

1. 請寄送 2 份發票。
2. 請按照上述價格、條款、交貨方式及項目輸入此筆訂單。
3. 若您無法按上述條件發貨，請立即通知我方。
4. 若延遲交貨，訂單將被取消。

填表人　　　　　　　　　　　　批准人
香農・桑德斯　　　　　　　　　羅素・C・哈珀

商業發票

出口日期 20××.03.13	參照號碼 發票編號 I-380033 航空貨運單號 710-8371-9011
銷售條款 CIF（成本保險費＋運費）	
託運人／出口商 拉斯克實業公司 澳洲 4707 艾希伯格市魏堡街 79 號	**收貨人** 有線網計算公司 美國 48001 密西根州馬斯基根蜜蜂街 2338 巷
貨物原產國 澳洲	**進口商** （若不同於收貨人）
最終目的國 美國	

貨物完整敘述	製造國	數量 （單位）	單價	總計
滑鼠矽膠帽（藍色）- 產品編號 0901	澳洲	5,500	4.50 美元	24,750.00 美元
小計		5,500		24,750.00 美元

包裹總數 55	運費	0.00
	保險	0.00

本商品、技術或軟體皆遵循澳洲出口管理條例。禁止違反澳洲法律的用途轉移。	**發票總額** 24,750.00 美元

本發票所登載之資訊皆為屬實，本批貨物皆與上述相同，以此證明。

託運人／出口商簽名

埃德加‧β‧隆巴爾多

日期
20××.03.13

文書範本 3 託運單 Bill of Lading

PSK 線　　　　　　　　　　　　　　　　　　　　　託運單

托運人 貝瑞貝瑞 美國 07102 紐澤西州紐瓦克 西區大道 4765 號 電話 +1-229-371-3091		裝箱單號 219476303-1	海運提單編號 SSOF0376181
		運費和應付費用支付方 托運人	
		銷售條款 FOB（裝運港船上交貨）	
收貨人 俄僻曲國際 英國 HU11 3NZ 斯普羅特利布 爾伍德路 49 號 電話 +44-814-391-9012		提單正本張數： 3	
		地點和簽發日期 美國紐瓦克／20×× 年 11 月 12 日	
通知方 同收貨人		如需發貨，請聯繫 普拉斯美機構	

出貨地 美國紐瓦克	裝貨港 美國紐瓦克	集裝箱數量 1	承運人收到的包裹數量 200
交貨地 英國樸茨茅斯	卸貨港 英國樸茨茅斯	轉運至 -	船名／航次 LSPL UBRE / 0038E

運輸標誌	包裹數量	包裹及貨物品名	總重量	測量
20' 鋼乾 貨櫃編號 POSL3084384	200 包	200 包野莓	900 公斤	11.1 立方公尺

運費已預付
裝運日期：20×× 年 11 月 14 日
承運人簽名：朗尼‧J‧埃克斯

裝箱單

發貨地 中心材料 美國 10007 紐約州紐約市泰勒街 3776 號	到貨地 英卓奈斯翠公司 南非共和國 3624 派恩敦車站路 1040 號 收件人：艾倫·蒙森

訂購日期 2012 年 10 月 11 日	訂單編號 967173	發貨日期 2012 年 10 月 21 日
運送業者 總統線	貨櫃編號 TA90198	發票號碼 BT-1923
包裹總數量 375	總重量 23,437.5 KGS	空運 / 提單編號 TXV7483503120

編號	品項編號	品名	數量	包裝類型	每件重量
1	Q18602	聚酯纖維	125	箱子	87.5KG
2	Q18605	聚丙烯	250	箱子	50.0KG
3					
4					

備註：本商品、技術或軟體皆遵循美國出口管理條例。

簽名：*康妮·D·阿吉拉爾*　　　　日期：20×× 年 10 月 21 日
康妮·D·阿吉拉爾，出口管理人

財務文件

 文書範本 1 信用狀 Letter of Credit

全州銀行

發行日期：20××10 月 15 日

不可撤銷信用證編號：9113814

開證行

全州銀行

10016 紐約州紐約市赫夫曼大道 3528 號

受益人	申請人
德瑞費特公司 55415 明尼蘇達州明尼亞波利斯威爾森街 204 號	賴瑞克斯集團 17815 賓夕法尼亞州布魯姆斯堡斯東尼朧森路 3124 號

金額： 五十七萬三千九百零二十分美元（573,900.20 美元）

到期日： 20×× 年 4 月 14 日

各位先生、女士：

應賴瑞克斯集團（「申請人」）的要求，本行將以此文件向您開出不可撤銷信用證，金額為五十七萬三千九百零二十分美元（US$573,900.20）。當您在本行上述的辦公室付款後，並向我們提供即期匯票及下列文件，即可取得本信用狀。

1. 受益人簽字聲明如下：

   德瑞費特公司證明賴瑞克斯集團未能完成項目 #AB391 的要求，因此我們有權獲得 573,900.20 美元的款項。

2. 受益人簽發給申請人的信件副本。

本行保證將按本信用狀各項條件開具匯票，於到期日或到期日前準時付款。

全州銀行

詹姆斯·W·克魯茲

授權簽名

詹姆斯·W·克魯茲，信貸經理

授權簽名者的姓名和職務

麥斯超市
資產明細表
截至 20×× 年 12 月 31 日的財政年度

資產	
流動資產	
現金及現金等價物	81,900,000
短期投資	-
存貨	59,833,000
其他流動資產	14,398,000
流動資產總計	**156,131,000**
長期投資	201,390,000
不動產、廠房及設備	593,090,000
無形資產	-
其他資產	76,010,000
資產總計	**1,026,621,000**

負債	
流動負債	
應付帳款	65,550,000
短／長期借款與債務	15,330,000
其他流動負債	1,300,000
流動負債總計	82,180,000
長期借款與債務	36,700,000
少數股東權益	2,900,000
其他負債	-
負債總計	121,780,000

股東權益	
優先股	-
普通股	16,900,000
留存收益	46,990,000
庫存股	5,550,000
資本盈餘	3,500,000
其他股東權益	-
股東權益總計	72,940,000
負債及股東權益總計	194,720,000

麥斯超市
損益表
會計年度截至 20×× 年 12 月 31 日

利潤	
銷售淨額	306,102,000
其他收入	1,900,000
總收入	308,002,000
銷貨成本	248,654,000
總收入	**59,348,000**

支出	
研究及發展費用	-
一般銷管費用	21,180,000
非經常性收入	-
其他	5,330,000
總支出	**26,510,000**
營業利益（或損失）	**32,838,000**

稅前利益	32,838,000
所得稅	4,712,000
計算少數股權前之收益	28,126,000
少數股東權益	230,000
淨利	**27,896,000**

諾瓦斯科   **日期：20××年10月5日**

薪資明細

員工姓名：賽斯‧克里斯普   員工編號：INT1091
職銜：財務顧問   支薪月份：20××年9月

收入		扣除額	
基本工資	1,600.00	所得稅	101.45
加班費	-	勞退	45.00
獎金	400.50	健保	23.00
醫療津貼	100.75		
交通津貼	200.00		
伙食津貼	110.00		
其他津貼	-		
總收入	**2,411.30**	**總扣除額**	**169.50**

淨工資		2,241.80

今年累計薪資	
應納稅所得額	20,801.70
所得稅	896.40
勞退	414.00

支付方式：電子轉帳
銀行：花旗銀行
帳號：0-163812-685

EZ TALK

向美國百大企業學商用英語

作　　者：SD語學研究所
譯　　者：阿譯
主　　編：潘亭軒
責任編輯：鄭雅方
封面設計：管仕豪
內頁設計：管仕豪
內頁排版：簡單瑛設
行銷企劃：張爾芸

發 行 人：洪祺祥
副總經理：洪偉傑
副總編輯：曹仲堯
法律顧問：建大法律事務所
財務顧問：高威會計師事務所
出　　版：日月文化出版股份有限公司
製　　作：EZ 叢書館
地　　址：臺北市信義路三段151號8樓
電　　話：(02)2708-5509
傳　　真：(02)2708-6157
客服信箱：service@heliopolis.com.tw
網　　址：www.heliopolis.com.tw
郵撥帳號：19716071日月文化出版股份有限公司

總 經 銷：聯合發行股份有限公司
電　　話：(02)2917-8022
傳　　真：(02)2915-7212
印　　刷：中原造像股份有限公司
初　　版：2023年8月
定　　價：430 元
I S B N：978-626-7329-23-8

向美國百大企業學商用英語 /SD 語學研究所著；
阿譯譯 . -- 初版 . -- 臺北市：日月文化出版股份
有限公司 , 2023.08
280 面 ; 19 X 25.7 公分 . -- (EZ Talk)
譯自 : 미국 100 대 기업 비즈니스 영어문서 작성법
ISBN 978-626-7329-23-8 (平裝)

1.CST: 商業書信　　2.CST: 商業英文
3.CST: 商業應用文　4.CST: 寫作法

493.6　　　　　　　　　　　112009237